Tempo and Mode in Evolution

A COLUMBIA CLASSIC IN EVOLUTION

Tempo and Mode in Evolution

GEORGE GAYLORD SIMPSON

COLUMBIA UNIVERSITY PRESS
NEW YORK

Library of Congress Cataloging in Publication Data

Simpson, George Gaylord, 1902–
   Tempo and mode in evolution.

   Reprint. Originally published: New York : Columbia
University Press, c1944.
   Includes bibliographies and index.
   1. Evolution.   I. Title.
QH371.S54   1984   575   83-23132
ISBN 0-231-05847-0

Columbia University Press
New York   Guildford, Surrey
Copyright © 1984 Columbia University Press
All rights reserved

Printed in the United States of America

p 10 9 8 7 6 5 4 3 2

Contents

TABLES	vii
FIGURES	ix
PREFACE	xi
INTRODUCTION: FORTY YEARS LATER	xiii
INTRODUCTION	xv

I. RATES OF EVOLUTION ... 3
    Relative Rates in Genetically Related Unit Characters ... 4
    Relative and Absolute Rates in Genetically Independent Characters ... 7
    Correlative Rates ... 12
    Organism Rates ... 15
    Group Rates and Survivorship ... 20

II. DETERMINANTS OF EVOLUTION ... 30
    Variability ... 30
    Mutation Rate ... 42
    Character of Mutations ... 48
    Length of Generations ... 62
    Size of Population ... 65
    Selection ... 74
        The Role of Selection ... 74
        Intensity of Selection ... 80
        Direction of Selection ... 83
    Summary ... 93

III. MICRO-EVOLUTION, MACRO-EVOLUTION, AND MEGA-EVOLUTION ... 97
    Minor Discontinuities of Record ... 99
    Major Systematic Discontinuities of Record ... 105
    Explanations of Systematic Discontinuities of Record ... 115

IV. LOW-RATE AND HIGH RATE LINES ... 125
    Distributions of Rates of Evolution ... 126
    Factors of Bradytely ... 135
    The Survival of the Unspecialized ... 142
    Relicts ... 144
    Summary ... 147

Contents

V.	INERTIA, TREND, AND MOMENTUM	149
	Rectilinear Evolution	150
	Evolutionary Trends in the Equidae	157
	Primary and Secondary Trends	164
	Evolutionary Momentum	170
	Theorems on Inertia in Evolution	177
VI.	ORGANISM AND ENVIRONMENT	180
	The Nature of Adaptation	180
	Real and Prospective Functions	183
	Preadaptation and Postadaptation	186
	Adaptive Zones	188
	The Adaptive Grid	191
VII.	MODES OF EVOLUTION	197
	Speciation	199
	Phyletic Evolution	202
	Quantum Evolution	206
	Summary	217
WORKS CITED		219
INDEX		227

Tables

1. Measurements on M^3 of five samples of fossil horses — 9
2. Correlation and regression with thickness of strata for two characters of *Kosmoceras* — 14
3. Rates of evolution in horses, chalicotheres, and ammonites in terms of number of genera per million years — 17
4. Estimates of changes in rates of evolution in genera of horses — 18
5. Estimates of durations of Tertiary epochs in millions of years — 19
6. Distribution of genera of Pelecypoda — 21
7. Distribution of genera of Carnivora (except Pinnipedia) — 22–23
8. Expected and actual generic survivorship in Pelecypoda and Carnivora — 27
9. Survival data for the Carnivora — 28
10. Statistical summary of representative data on carnivore evolution — 29
11. Coefficients of variation for paracone height and ectoloph length in four species of horses — 38
12. Genotypes in successive generations — 41
13. Occurrence of a cingulum on lower cheek teeth of *Litolestes notissimus* — 60
14. Occurrences in America of Apatemyidae and early Equidae — 73
15. Available records of ancestry of the mammalian orders — 108–109
16. Estimated durations of the orders of mammals — 120
17. Expected and realized survival into the recent of broad Pelecypod genera of various ages — 130
18. Progressive curvature in the shell from a primitive *Ostrea* to an advanced *Gryphaea* — 171
19. Characteristics of the main modes of evolution — 217

Figures

1. Relative rates in horses — 5
2. Relationships of structural changes in two characters of five genera of Equidae — 10
3. Relative rates of evolution in equid molars — 10
4. Phylogeny of five genera of Equidae and evolution of two related tooth characters — 11
5. Survivorship curves for genera of pelecypods and of land carnivores — 25
6. Survivorship in pelecypod genera, land carnivore genera, and *Drosophila* individuals — 26
7. Variation in a branching phylogeny — 34
8. Variability, range, genetic structure, and adaptability — 36
9. Continuous and discontinuous phenotypic variation in a fossil mammal — 43
10. Selection vectors — 84
11. Selection landscapes — 90
12. Two patterns of phyletic dichotomy — 91
13. Major features of equid phylogeny and taxonomy represented as the movement of populations on a dynamic selection landscape — 92
14. An apparent saltation in an ammonite phylum and its interpretation as caused by a depositional hiatus in the beds from which specimens were collected — 100
15. Apparent saltatory and true continuous phylogeny of the Equidae — 104
16. Systematic deficiencies of record of mammalian orders — 107
17. Reduction of number of individuals demonstrating major structural changes — 118
18. Frequency distributions of rates of evolution in genera of pelecypods and land carnivores — 128
19. Realized and expected age compositions of recent pelecypod and land carnivore faunas — 131
20. Bradytely in pelecypods — 132

Figures

21. Survival of the unspecialized in Caenolestoidea — 142
22. Example of characteristic changes in distribution of a group of animals in the whole course of its history — 146
23. Analysis of some supposed momentum effects in evolution — 175
24. Diagram of the realization of functions by overlaps of prospective functions of organisms and environment — 184
25. Convergence and divergence in evolution of phylogenetically prospective functions of organisms through overlap with the prospective functions of their environments — 185
26. Diagram suggesting the complexity and nature of adaptive grid — 191
27. Diagrammatic representation on the adaptive grid of conditions leading to survival — 192
28. Diagrammatic representation on the adaptive grid of a bradytelic group, a tachytelic line arising from it, and the subsequent deployment and further evolution of this line as a horotelic group — 193
29. Diagrammatic representation on the adaptive grid of the step-like evolution of a group through successive occupation of different adaptive zones — 194
30. Greatly simplified representation on the adaptive grid of the evolutionary history of the Felidae — 195
31. Diagrams of characteristic examples of the three major modes of evolution — 198
32. Two patterns of speciation — 200
33. Three patterns of phyletic evolution — 204
34. Phases of equid history interpreted as quantum evolution (see also Fig. 13) — 208
35. Diagram of "explosive" evolution by multiple quantum steps into varied adaptive zones, followed by extinction of unstable intermediate types and phyletic evolution in each zone — 213
36. Intergroup variation induced by the onset of unfavorable conditions — 215

Preface

THE MANUSCRIPT of this work was begun in the spring of 1938. Its composition was therefore not hasty, but the final revision was made under conditions of stress, and not all passages are to be taken as of the same date. Some important studies that were not at hand until about the time when circumstances forced me to discontinue revision were necessarily omitted from consideration. Some parts, written several years ago, could advantageously have been altered, had I been given more leisure, but they are believed to be valid as they stand.

Because of the exigencies of military service abroad I have been unable to concern myself with this work since the manuscript was completed, and I am indebted to my wife, Dr. Anne Roe, to Dr. Edwin H. Colbert, and to Miss Ida Lynn for preparing the final manuscript for publication and supervising its production in my absence. The figures were prepared (from rough sketches by me) under the supervision of John C. Germann.

GEORGE GAYLORD SIMPSON

New York
April, 1944

Introduction: Forty Years Later

THE BOOK NOW BEFORE YOU was written intermittently from 1938 to early 1942, when its writing was interrupted by overseas service in the army. As noted in the preface, its final preparation, not requiring rewriting, was completed in my absence by my wife, some staff members at the American Museum of Natural History, and the staff of the Columbia University Press. The present edition is thus being published more than forty years after the book was written. Except for the addition of this new foreword, the book is here reprinted without any change from the first printing. The purpose of this new foreword is to outline briefly some antecedents of the book, its place in the study of evolution as a whole, its effects on such studies as far as the author can judge, and something about developments not foreseen or foreseeable forty years ago.

In 1894 a Columbia University Biological Series of books was started under the editorship of Henry Fairfield Osborn, then a professor at the university. Early volumes under that sponsorship and editorship were published by the Macmillan Company, as Columbia did not then have publishing facilities of its own, only an editorial board. The first volume in the series was *From the Greeks to Darwin*, written by the editor, Osborn, on the basis of lectures given at Princeton and at Columbia. Although Osborn was a paleontologist, this volume is concerned with the history, not particularly paleontological, of the concept of evolution.

As its title indicates, the first book in the Columbia Biological Series was devoted to the thesis that evolutionary concepts, both philosophical and biological, themselves have evolved from the ancient Greeks, many of whom Osborn specified from Thales (born in 624 B.C. according to Osborn, more probably about 636 B.C.) to Galen (born A.D. 131 according to Osborn, somewhat dubious but probably close). Traced on through the centuries, the concept is seen as reaching its full development by Darwin and by Wallace, whom Osborn considered essentially post-Darwinian. (Wallace was still living and writing when Osborn wrote this book.) Osborn ended his book with the statement that only the future can determine whether (as of 1894) the "old, old problem" (of the causes of evolution) had been fully answered or "whether we should look for still another Newton in our philosophy of Nature." Historians of biology, more pertinent here than the majority of historians of science, who usually have stressed the physical sciences or confined themselves to them, have modified or rejected Osborn's thesis of a continuous evolution of evolutionary thought from Thales to Darwin, or Wallace, or to Osborn himself. I agree with what I think is the present consensus of historically minded biologists or biologically minded historians. Lamarck's theory was not supported by objective evidence, and like all scientists he inherited a body of facts, conjectures, and theories, but Lamarck was the first to be explicitly and completely committed to an evolutionary concept of the history of life.

Not only after Lamarck but even after Darwin (whom Osborn had met when Darwin was old and Osborn young), Osborn said, as previously noted, that another Newton was needed to solve the problem of an evolutionary philosophy of nature. In later years it seemed probable that Osborn had come to think of himself as the other Newton. As a continuation of what he thought of as a tradition continuous from the early Greeks, he sought a more explanatory theory of evolution. He developed this in a series of papers, especially in the 1910s and 1920s, and in the two massive memoirs, on titanotheres (1926) and proboscideans (published posthumously in 1936 and 1942), to which he devoted much of the last decades of his life, beginning as early as 1910. (He died at the age of 78 years on November 6, 1935.) His final views were especially set out in a series of papers devoted (under this title and some others) to "The Origin of Species as Revealed by Vertebrate Paleontology." The gist of his theory was first called "rectigradation" and later "aristogenesis." This was defined as "the gradual, secular, continuous, direct, reactive, adaptive origins of new biomechanisms" in what he first called the "germ-plasm" and later the "geneplasm," which did not refer to genes in the now usual genetical sense. (The word "gene" had already been used by T. H. Morgan in its present sense in 1917.) As to the modes of the origin of biomechanisms, Osborn concluded that "We are now on absolutely sure ground. This ground is contra-Lamarckian and contra-Darwinian. It is also contrary to the neo-Darwinian evolutionary hypotheses of the leading biologists and geneticists of our day."

The "absolutely sure ground" was paleontological observations and deductions, especially those Osborn made from his studies of titanotheres and proboscideans, among other fossil mammals. In his final years, Osborn was aware that his theory was not accepted by most paleontologists or by other biologists. He consoled himself with the belief that it would be accepted sooner or later. It has not been.

Osborn obviously was not the first paleontologist to relate the fossil record to evolutionary theory. Darwin was a geologist and paleontologist before he overtly espoused evolution, and two chapters in *The Origin of Species* are devoted primarily to paleontology. Apart from that, Edward Drinker Cope (1840–1897), personally known to Osborn early in Osborn's career and in the last years of Cope's, had been a convinced evolutionist. He published two books mainly devoted to paleontology and evolution: *The Origin of the Fittest* (1886) and *Primary Factors of Evolution* (1896). In his long biography *Cope: Master Naturalist* (1931) Osborn stigmatized Cope for being a Neo-Lamarckian and for inventing new terms such as "catagenesis." There is irony in this, for despite his claim to be "contra-Lamarckism" there was a strong Lamarckian element in Osborn's theory, and he was also prone to invent new terms, like "aristogenesis."

All this is relevant, at least indirectly, to the book here reprinted with this new foreword. There is also a curious connection. At the opening of a new building for the Peabody Museum at Yale University in December 1925 Osborn gave an address in which he said; "Perhaps within the very walls of the Peabody Museum, where adaptation is set forth so transparently by the master hand of Lull, some young Aristotle or Darwin may find his inspiration to grasp the problem of the origin of species which has baffled man for two thousand five hundred and eighty-five years." (He must have been counting from 660 B.C., which would have started the search even before Thales.)

It is quite obvious that I am not a Darwin, and still less an Aristotle, but it happened that a bit later in the year (1926) when that statement of Osborn's was published I became a Ph.D. after studying in the Yale Peabody Museum under the tutelage of Richard Swann Lull (1867–1957). Lull was another of the then increasingly numerous paleontologists interested in evolutionary theory. He was Neo-Darwinian in general, especially as regards the reality and importance of natural selection, but he tended to give almost equal time to opposing or marginal opinions that he considered unlikely but not impossible. Not much later, from 1927 to 1935, I was closely associated with Osborn at the American Museum of Natural History.

Before joining the American Museum's curatorial staff, I had published twenty-five papers and finished writing two large memoirs on Mesozoic mammals later published as hardback books by the British Museum (Natural History) (1928) and Yale University Press (1929). Obviously these are now outdated, but they are still found useful and have recently (1980) been reprinted together in a single volume. On the American Museum staff I had on hand an even greater collection of fossils than at the Yale Peabody Museum, and I was launched into more varied descriptive and taxonomic studies mainly of fossil mammals. I was also almost immediately involved in field work, finding and collecting fossil vertebrates, especially mammals. Even on the basis of what I then knew, I was firmly convinced that evolution is a fact, obvious to anyone really acquainted with the evidence. Of course I have remained so ever since, all supposed evidence to the contrary being absurd or, less commonly but sometimes, simple prevarication.

As a result of training and experience, I also habitually thought in four dimensions, time being the fourth and being particularly paleontological. I was sometimes astonished to find that nonpaleontological biologists did not all think in this dimension as well as the physical three. I had read *The Origin of Species* and some others of Darwin's books and many papers on evolution, as well as straight paleontology, in English, French, and German, the latter two, in that order, having been required for my Yale degrees. I also had the advantage of having spent most of a field season (summer 1924) alone with William Diller Matthew (1871–1930). He was a great paleomammalogist, a hero to me, and with him familiarity bred only respect and admiration.

Of course I also read most of the work of Osborn, whom I thought and still think of as "The Professor." I respected and admired him too, but admittedly in a somewhat lesser way. As his theoretical views were developed while I knew him, they seemed to me to become more vague and his "absolutely sure ground" in paleomammalogy to become rather a quagmire. He knew that I disagreed, but he did not resent disagreement if it was courteous, and he probably expected that I would come around to his views as I learned more. I was intensely interested in evolutionary theory, but for about my first ten years there I did not think I knew enough to judge extant theories well or perhaps even to add something to them. I was inclined toward Neo-Darwinism to the extent of considering natural selection as the principal but not necessarily the only nonrandom or directive element in evolution. In fact I was rather dismayed that some of the members of the Museum's scientific staff were anti-selectionists. They were also anti-Neo-Lamarckian and anti-Osbornian in their theoretical views.

I was not Neo-Lamarckian, as I was convinced that acquired characters are not inherited, and also that there is no *scala naturae,* which had been the other main Lamarckian principle but was not Neo-Lamarckian. I was also not a Neo-Darwinian in a strict sense of the term. Of course I knew that Darwin had considered the inheritance of acquired characters as a subsidiary factor in evolution, and that most evolutionary theorists before him, including Lamarck and Charles Darwin's grandfather Erasmus, and even some of my own contemporaries considered this a major or *the* major factor in evolution. It had by my time become clear that Darwin's hypothesis of how this occurred, even if it occurred, was flatly wrong. I did not believe that this occurred in any way, as the few remaining Neo-Lamarckians did. I also had learned while still in college that a whole new biological science of genetics had been developed since 1900.

Darwin also believed that what he called "sports" and geneticists were later calling mutations with large somatic effects might have some but probably a minor role in evolution. I was inclined to be dubious, but felt unable to decide. As all biologists now know, 1900 was the approximate date of what some geneticists did and do consider even now the "rediscovery" of Mendel's laws and thus of what they call Mendelism. Even before I wrote *Tempo and Mode in Evolution* I was somewhat put off by this. I did not doubt that the "laws," derived by Mendel from experimentation (perhaps with a little fiddling of his statistics) do apply to some aspects of inheritance in organisms. What I considered historically wrong was the opinion that Darwin was at fault in not being aware of these "laws," published in 1866, after the first edition of *The Origin of Species* (1859) but well before Darwin's last revision, the sixth, published in 1872. I also was more than dubious about the direction taken by many geneticists after 1900. As this is involved in leading up to my writing of *Tempo and Mode in Evolution,* it requires at least brief further notice here.

The usual explanation of the tardiness of Darwin and other biologists to follow up Mendel's lead is that his crucial paper, in German, was not distributed widely enough for general notice. Such is not true. Mendel's paper was published in *Verhandlungen naturforschender Verein in Brünn* (Proceedings of the Naturalists' Union in Brünn). It was regularly distributed rather widely, and Mendel also distributed some copies. It would almost inevitably have interested any evolutionist especially interested in heredity, which Darwin surely was. There is no evidence one way or the other on whether Darwin did ever read or even hear of Mendel's paper. My point is that there was no particular reason why Darwin should have done so. Mendel made it clear that he was an anti-evolutionist, of which there were still all too many. To Mendel, his experiments showed how a population of a species could be quite variable but that species or crosses among subspecies or variants could be stable *without* evolving. Darwin's point was that species have evolved and that variants or subspecies can become species separate from their ancestry. That populations of species usually have evident variation was already well known to Darwin and clearly stated by him. Thus from Darwin's point of view, even if he had known of Mendel's conclusions, they would have been wrong on one main point and banal or superfluous on the other.

The "rediscoverers" of "Mendelism" were Hugo De Vries, Karl Erich Correns, and Erich Tschermak von Seysenegg. De Vries experimented with a species of evening primroses, *Oenothera lamarckiana,* and found that among them appeared characters strikingly different from those of either of their parent plants. He called these large and sudden genetic changes "mutations" and proposed a mutation theory of evolution on this basis. (Parenthetically I note that the word "mutation," meaning any sort of change, is old in English, having been used with a different spelling as early as Chaucer. It had also been used before De Vries in a technical sense by paleontologists for a perceptible change through geological time within an evolving lineage. Paleontologists have not used the word in that sense since it has been taken over by the geneticists, who in the time of De Vries were unaware of the former paleontological usage.) According to De Vries' mutation theory evolution has been more effected by mutations, as he defined them (Darwin would have called them "sports"), than by the lesser variants of Darwinian theory. In that conception, the dominant process in evolution is chance mutation and not natural selection.

While I was still in graduate school it had been found, especially by O. Renner, that *Oenothera lamarckiana* is a peculiar organism and that its De Vriesian "mutations" comprise a mixture of different sorts of heredity likely to occur very rarely in nature. Then for a time geneticists thought (incorrectly as it has turned out) that all changes in heredity are of the same sort and can range from barely perceptible to Darwinian "sports" or De Vriesian "mutants." Nevertheless some continued to believe that mutation in that more general sense has been more effective in evolution than is natural selection, and that adaptation is largely by chance. When I joined the American Museum staff I was surprised to find that some very good zoologists, such as G. Kingsley Noble, a distinguished student of living amphibians, held that view.

In 1936 Harvard University celebrated its tercentenary with "a conference of arts and sciences," and I was invited to attend and to give an address. Although I was then 34 years old and had written more than 150 publications, this was my first strictly theoretical paper on an aspect of evolution as seen in the fossil record. I felt it rather daring of me. The paper was titled "Patterns of Phyletic Evolution" and was published in 1937 in the *Bulletin of the Geological Society of America*. In 1936 the American Society of Naturalists, with several other more specialized societies, held a symposium on "Superspecific Variation in Nature and in Classification" at a meeting with the American Society for the Advancement of Science. I read a paper on that subject "From the Point of View of Paleontology." The first part of this was based on studies of fossil mammals that I had collected in Patagonia (southern Argentina) and was then monographing, and the second part was a criticism of Kinsey's views based on his massive study of the wasp genus *Cynips*, then recently published. (Kinsey also contributed to the symposium; this was before his concentration on human sex.) My paper was published in *The*

American Naturalist in 1937. (I have been criticized, especially by Ernst Mayr, for not discussing the origin of species by cladistic separation of evolving populations in *Tempo and Mode in Evolution*.) In it, one of my earliest theoretical papers, I stressed and diagrammatically illustrated the fact that "super-specific" taxa necessarily arise on a basis of such speciation. You will find this also present in the book here reprinted.

In 1939 there was published a book on the application of statistical methods to living and fossil mammals, entitled *Quantitative Zoology* and jointly written with my childhood friend Anne Roe, who used statistical methods in her research in psychology, and who became my wife in 1938. In 1941, while I was writing the present book, I had published two theoretical papers and an abstract relevant to it: "The Role of the Individual in Evolution," "Quantum Effects in Evolution" (an abstract), and "Range as a Zoological Character." The latter was statistical, and it as well as *Quantitative Zoology* are relevant here because they represent the analysis and instrumentation of a concept of species and higher taxa as populations and not, as was long customary (and sometimes is held even now) as a sort of extended abstraction of taxa as individuals.

I have in a very summary way discussed some of the things that led up historically and in my own experience to the writing of the book here before you. Now I turn to the series of books on evolution in which this book first appeared.

Columbia Biological Series books published by Macmillan following the first, by Osborn, did not again deal with any particular approach, historical or otherwise, to evolution. The series was discontinued for a time, but it was taken up again, now published by Columbia University Press. The general editor for this revised series was Leslie Clarence Dunn (Dunny to his friends), a geneticist then a professor in the Department of Zoology at Columbia. Under his editorship four books on evolution were published from 1937 to 1950. The first, which was the eleventh in the series as a whole from 1894 on, was by Theodosius Dobzhansky (Dobie to his friends), another geneticist, then also a professor in the Department of Zoology at Columbia. This book was *Genetics and the Origin of Species,* published in 1937, with lightly revised new editions in 1941 and again in 1951. Second was *Systematics and the Origin of Species* by Ernst

Walter Mayr (Ernst to his friends), then associate curator of birds in the American Museum of Natural History. This was published in 1941 as thirteenth in the Biological Series. It was not followed by a revised edition as such. Third was by me (George to some friends, G to others) and was the book now before you as a reprint. It was first published in 1944 as the fifteenth book in the Biological Series. As will be noted later, it was followed in 1953 by a completely rewritten and retitled version, also published by Columbia University Press. The last of the four books devoted to evolution in the period here under consideration was *Variation and Evolution in Plants* by George Ledyard Stebbins, Jr. (Ledyard to his friends), a botanist and a professor of genetics at the University of California at Berkeley when he wrote the book. (He transferred to the campus at Davis the year it was published.) It was number sixteen in the Columbia Biological Series, published in 1950, and not followed by a revision as such.

In his introduction to the first of those four books, which was also the first of the resuscitated Series, Dunn wrote: "There was need for such a summary and synthesis of the new experimental evidence, and for reassessment of the older theories."

Although they had predecessors, it has been generally considered that these four books, considered together, established an approach to evolutionary theory that was and still is a synthesis as suggested by Dunn as editor. This approach has been generally known as a, or the, "synthetic theory," because it brings together and in a sense coordinates results of all the many specialized subsciences that bear on evolution. (Rather oddly, Dobzhansky mildly objected to the designation "synthetic" because he thought that some people might think of anything "synthetic" as artificial or not genuine!) Especially in Europe it is sometimes called "Neo-Darwinian" because it accepts Darwinian natural selection as a—not necessarily *the*—nonrandom feature of evolution. That designation is, however, historically and materially wrong. One has only to read Dobzhansky's own book in this series to see that the synthesis then already went far beyond and even contrary to Neo-Darwinism as so designated in the late nineteenth century.

It is unnecessary here to say why I wrote *Tempo and Mode in Evolution* because I think that is made perfectly clear in the original introduction, also reprinted here. The way in which I approached this task is perhaps not quite so clear there, but is evident throughout the book. It will also be evident that I had studied and profited by Dobzhansky's *Genetics and the Origin of Species,* the first edition of which was published not long before I started to write this book and the second (first revised) edition while I was writing it. It is apparent also that when I wrote I had not read Mayr's *Systematics and the Origin of Species.* That was published before *Tempo and Mode in Evolution* but after I had finished writing it and at a time when I could not study, and in fact could not obtain, a copy of Mayr's book.

This book has sometimes been dismissed as devoted only to the proposition that paleontology is not contradictory to genetics as genetics was at the time I wrote. That was one of my aims, but I do not see how anyone who has really read this book could fail to understand that it was not my only or even my main aim. My main aim was to explore and in a way to exploit the fact that paleontology is the only four-dimensional biological science: time, "tempo," is inherent in it. Thus the aim of this book, which I think it accomplished, was to bring this dimension squarely, methodologically, into the study of evolutionary theory.

The pre-Darwinian English classics of creationism were the Bridgewater Treatises and Paley's "Natural Theology." These emphasized the obvious fact that all organisms are adapted to live where, when, and how they do live. Their conclusion was that this can be explained only by a Creator following up a divine plan. The problem and the strategy for Darwin was therefore to find some way in which evolution, without divine creation as such, could produce the observed results of adaptation. He found it in natural selection. From 1900 on the followers of the new science of genetics had tended to oppose the theory of natural selection. Although he had forerunners, whom he cited, Dobzhansky's book in this series both by field and by laboratory studies established the recognition that natural selection can and does produce adaptive evolution.

In the book here present I adopted that theory and exemplified it as a process occurring in the time dimension. In this respect it will also be noted that I discussed the views of Goldschmidt, an able naturalist and in an odd way a geneticist who did not think in terms of genes. He maintained that marked adaptive changes in evolution occurred not by natural selection in the course of generations but by the chance production of "hopeful monsters" by what he called "systemic mutation," instantaneous remodeling of the whole genetic system. Early in the present century some able paleontologists, notably William Diller Matthew (1871–1930) also found evidence for natural selection in the fossil record. However in 1936 another able paleontologist had attempted to reconcile the different views of early geneticists with those of paleontologists but had come to a conclusion quite like Goldschmidt's although different in detail or mechanism. In the present book I opposed both Goldschmidtian genetics and Schindewolfian paleontology for reasons fairly clear in the following pages.

Between 1944 and 1953 I found myself a *homo unius libri*, in spite of my book *The Meaning of Evolution* having been published by Yale University Press in 1949. However, that book was meant to be, and was, a popular book, widely read and translated into many languages but rarely noticed by my more technical colleagues. Also in 1949 was published by Princeton University Press a symposial volume, *Genetics, Paleontology, and Evolution,* edited by Glenn Jepsen, me, and Ernst Mayr, with a foreword by Jepsen and chapters by me and by Mayr. This was oriented by the synthetic theory, which was also having wide influence internationally on all students of evolutionary theory. By 1951 so much more had been done in this field that, as I remarked, *"Tempo and Mode in Evolution* [had] helped to produce its own obsolescence," and that "[It] had served its purpose and should be allowed to fossilize quietly." However Columbia University Press said it was not dead enough for that and wanted it to be revised instead. I revised it so thoroughly that although it followed the plan of *Tempo and Mode* it was completely rewritten and considerably lengthened (from 237 to 434 pages). The manuscript was completed at the end of 1951 but editing, indexing, proofreading, and manufacture carried publication by Columbia University Press into 1953.

Because of the changes, additions, and rewriting that book was given a different title, *Major Features of Evolution.* In general I continued to be *homo unius libri.* In my case the one book cited continued to be *Tempo and Mode* in some studies by others, but more commonly it has been *Major Features*. One or two evolutionists who had welcomed *Tempo and Mode* as a good pioneering try nevertheless have condemned *Major Features* as not equally pioneering and even as being an effort to fix a dogma, which indeed is not true as those who have read and understood it have usually recognized. Some whose attention was on speciation and minor features of evolution have also failed to appreciate that the intention of the book was to elucidate *major* features, which it did.

In this thorough revision I had finally at hand Mayr's book in the Columbia Biological Series, the second revision (third edition) of Dobzhansky's book in the series, and Stebbins' book in the series. There were also a large number of other relevant new studies, almost all in the general trend of the synthetic theory more or less as set out in the four books of the series including *Tempo and Mode* on the paleontological side. An early exception was Schindewolf, who in 1950 had published two more works continuing his view that his paleontological data supported evolution by "mutations" like those of De Vries or like Goldschmidt's "systemic mutations." In *Major Features* somewhat more adverse critical attention was therefore given to Schindewolf's views. It may interest the reader that Schindewolf subsequently visited the United States and the American Museum of Natural History, where I was then working, and that he carefully avoided me. It is also of interest that in 1958 Marjorie Grene, a classical philosopher, without firsthand knowledge of paleontology or genetics, wrote a long study of "Two Evolutionary Theories," mine and Schindewolf's. She decided in favor of Schindewolf mainly on the grounds that his view was closer than mine to those of the ancient Greeks. Francisco Ayala, who is the unusual combination of being a well-trained and active student both in philosophy and in evolutionary genetics, has demolished Grene's arguments on this subject, both philosophical and biological.

Before writing *Major Features* and thus in a sense rewriting *Tempo and Mode* I had also gone on with relevant theoretical studies as well as adding significantly to knowledge of the fossil record. Among other publications may be mentioned "The Problem of Plan and Purpose in Nature" (1947, in *The Scientific Monthly*), "Rates of Evolution in Animals" (a chapter in the 1949 book edited by Jepsen, me, and Mayr, noted above), a revision of *The Meaning of Evolution* (1967), "Evolutionary Determinism and the Fossil Record" (1950, in *The Scientific Monthly*), "Periodicity in Vertebrate Evolution" (1952, in *Journal of Paleontology*) and, among books, *This View of Life* (Harcourt, Brace and World, 1964).

All of the authors of what may be fairly considered the four classics of synthetic theory in the Columbia University Biological Series went on to expand and update their own views and also to bring other branches of science into the synthesis. Among Dobzhansky's later books is *Genetics of the Evolutionary Process* (Columbia University Press, 1970). Among Mayr's is *Animal Species and Evolution* (Belknap Press of Harvard University Press, 1963). Among mine the most recent is *Fossils and the History of Life* (W. H. Freeman/Scientific American, 1983). Among Stebbins' is *Processes of Organic Evolution* (Prentice-Hall, 1966). As an example of integration with other branches there is *Behavior and Evolution,* edited and with chapters by Anne Roe and me (Yale University Press, 1958).

The synthetic theory of evolution has never been supported as dogmatic or as an acceptable a priori approach to evolution in any publication that I know of. Knowledge is constantly advancing in all the relevant sciences, from molecular biology to ethology, to name two specialties with little in common except through evolutionary theory. The synthetic theory thus becomes more complex and modified in detail as time goes on, but it has not been clearly and indubitably contradicted as indeed a synthesis of everything known about evolution at a given time. So positive a statement requires brief notice of what does claim to be a "new and general evolutionary theory" which nominally rejects the synthetic theory. This is called "punctuated equilibrium," and the synthetic theory is condemned as "gradualistic." By "punctuation" is meant essentially what is meant by "quantum evolution" in *Tempo and Mode in Evolution*, the origin of a species or other taxon by exceptionally rapid evolution. In his first statement Gould called this "the Goldschmidt break," and he suggested that speciation may involve Goldschmidtian "rapid reorganization of the genome, perhaps non-adaptive." By "equilibrium" he meant that after the "Goldschmidt break" species usually do not change further for a considerable length of geological time. More recently Gould has essentially abandoned the "Goldschmidt" break and has redefined "punctuation" as being continuous but so rapid as to seem practically instantaneous in stratigraphic observation. He also has urged to take as given the fossil record, in which species and other taxa often, but not invariably, do appear rapidly or even "instantly," that is, in a span of geological time so short as not to have been measured. As all paleontologists have always known, the incompleteness of the fossil record is factual and is the most logical explanation of apparent "quantum evolution" or "punctuation." New species are still being found and named every day so the record becomes more and more reliable, but it is also obvious that fossils representing every species that ever lived do not now exist; thus, rich as the known record is becoming, it can never be literally complete. Nevertheless, as in *Tempo and Mode* and later studies, the fossil record does provide a factual basis for evolutionary theory.

Gould lumped together Darwinism, Neo-Darwinism, and Synthesis as "gradualism." As he defined "gradualism," it was a straw man for his attack. That term is not a description of any of the schools of thought that he wants to reject. "Gradualism" *sensu* Gould is one end of a continuum and "punctuated equilibrium" is the other end. We thus have here an "either-or" proposition, a Hegelian or Marxian dialectic. It is apt to point out that apparent contradiction between *thesis* and *antithesis* leads logically not to one or the other but to *synthesis*. That is what the synthetic theory does.

In closing it may be pointed out that the book you are about to read has been read by almost everyone who has been involved in the study of evolution since 1944. It has been widely circulated not only in English but also in French, German, and Russian.

G. G. Simpson

Introduction

THE BASIC PROBLEMS of evolution are so broad that they cannot hopefully be attacked from the point of view of a single scientific discipline. Synthesis has become both more necessary and more difficult as evolutionary studies have become more diffuse and more specialized. Knowing more and more about less and less may mean that relationships are lost and that the grand pattern and great processes of life are overlooked. The topics treated in the present study do not embrace the whole subject of evolution, but they are fundamental in nature and broad in scope. They are among the basic evolutionary phenomena that have tended to be obscured by increasing specialization and that overlap many different fields of research. Data and theories from paleontology, genetics, neozoology, zoogeography, ecology, and several other specialties are all pertinent to these themes. The complete impossibility of attaining equal competence and authority in all these fields entails unavoidable shortcomings, but the effort to achieve such a synthesis is so manifestly desirable that no apology is in order. The intention will hardly be criticized, whatever is said about its execution.

The attempted synthesis of paleontology and genetics, an essential part of the present study, may be particularly surprising and possibly hazardous. Not long ago paleontologists felt that a geneticist was a person who shut himself in a room, pulled down the shades, watched small flies disporting themselves in milk bottles, and thought that he was studying nature. A pursuit so removed from the realities of life, they said, had no significance for the true biologist. On the other hand, the geneticists said that paleontology had no further contributions to make to biology, that its only point had been the completed demonstration of the truth of evolution, and that it was a subject too purely descriptive to merit the name "science." The paleontologist, they believed, is like a man who undertakes to study the principles of the internal combustion engine by standing on a street corner and watching the motor cars whiz by.

Now paleontologists and geneticists are learning tolerance for each other, if not understanding. As a paleontologist, I confess to inadequate knowledge of genetics, and I have not met a geneticist who has demonstrated much grasp of my subject; but at least we have come to realize that we do have problems in common and to hope that difficulties encountered in each separate type of research may be resolved or alleviated by the discoveries of the other.

Earlier work in genetics was devoted mainly to studying the transmission of inherited characters, cryptogenetics. Earlier work in paleontology was devoted largely to the determination of the forms and sequence of fossil animals and to their classification. There is hardly any point of contact between these subjects, and it is not surprising that workers in the two fields viewed each other with distrust and sometimes with the scorn of ignorance. Now this basic work in both subjects, although far from completed in either, is so well advanced that what remains must in great part follow charted paths. Both geneticists and paleontologists are turning to new fields. In genetics the vigorous and trail-blazing work is now more likely to be in the field of phenogenetics—not how heredity occurs, but how hereditary characters achieve their material expression in the life cycle of an animal or plant and in the third major division of modern genetics, that of population genetics.

In paleontology there is fresh interest in attempting to infer not only the course but also the mechanisms of evolution. This is not new, to be sure, for even the earliest paleontologists did speculate as to causes, just as the early experimentalists did pay some attention to natural populations; but the attack is now being made with more hope and with better techniques. The paleontologist is acquiring a different attitude toward variation, which was only a nuisance to the classifier, but is now becoming an important study in itself and one that is improving greatly in method. Like the geneticist, the paleontologist is learning to think in terms of populations rather than of individuals and is beginning to work on the meaning of changes in populations. Thus, from very different starting points geneticists and paleontologists have come to the study of problems that are not only related but also sometimes identical.

For the study of these problems it is the great defect of paleontology that it cannot directly determine any of the cryptogenetic factors that must, after all, be instrumental in the evolution of populations. Fossil animals cannot be brought into the laboratory for the experimental determination of their genetic constitutions. The experiments have been done by nature without controls and under conditions too com-

Introduction

plex and variable for sure and simple analysis. The paleontologist is given only phenotypes, and attempts to relate these to genotypes have so far had little success. But here genetics can provide him with the essential facts. One cannot directly study heredity in fossils, but one can assume that some, if not all, of its mechanisms were the same as those revealed by recent organisms in the laboratory. One cannot identify any particular set of alleles in fossils, but one can recognize phenomena that are comparable with those caused by alleles under experimental conditions.

On the other hand, experimental biology in general and genetics in particular have the grave defect that they cannot reproduce the vast and complex horizontal extent of the natural environment and, particularly, the immense span of time in which population changes really occur. They may reveal what happens to a hundred rats in the course of ten years under fixed and simple conditions, but not what happened to a billion rats in the course of ten million years under the fluctuating conditions of earth history. Obviously, the latter problem is much more important. The work of geneticists on phenogenetics and still more on population genetics is almost meaningless unless it does have a bearing in this broader scene. Some students, not particularly paleontologists, conclude that it does not, that the phenomena revealed by experimental studies are relatively insignificant in evolution as a whole, that major problems cannot now be studied at all in the laboratory, and that macro-evolution differs qualitatively as well as quantitatively from the micro-evolution of the experimentalist. Here the geneticist must turn to the paleontologist, for only the paleontologist can hope to learn whether the principles determined in the laboratory are indeed valid in the larger field, whether additional principles must be invoked and, if so, what they are.

On two topics, in particular, the paleontologist enjoys special advantages. It is with reference to these topics that the geneticist and the general student of evolution now turn most frequently to paleontology for enlightenment and too often turn in vain. The first of these general topics has to do with evolutionary rates under natural conditions, the measurement and interpretation of rates, their acceleration and deceleration, the conditions of exceptionally slow or rapid evolutions, and phenomena suggestive of inertia and momentum. In the present study all these problems are meant to be suggested by the word "tempo." The group of related problems implied by the word "mode" involves the study of the way, manner, or pattern of evolution, a study in which tempo is a basic factor, but which embraces considerably more than tempo. The purpose is to determine how populations became genetically and morphologically differentiated, to see how they passed from one way of living to another or failed to do so, to examine the figurative outline of the stream of life and the circumstances surrounding each characteristic element in that pattern.

Readers have been led to expect that a study of evolution by a paleontologist will have as its subject matter, first, descriptions of the morphology and phylogeny of particular lines of extinct animals and, second, discussion of some empirical principles of morphogenesis such as convergence, irreversibility, or polyisomerism. The reader is warned at the outset that this study mentions neither of these subjects except incidentally. Phylogenetic examples are introduced as evidence and to give reality to the theoretical discussion, but they are not expounded in detail or as an end in themselves. Some of the so-called paleontological laws of morphogenesis are mentioned, but only as they bear on the broader problems of tempo and mode. Phylogeny and morphogenesis continue to be chief aims of paleontological research, but the present purpose is to discuss the "how" and—as nearly as the mystery can be approached—the "why" of evolution, not the "what." This is now a more stimulating point of view for the paleontologist, one more suggestive of new lines of study, and it is more immediately interesting to the nonpaleontological evolutionist who wishes to have the evidence interpreted in ways more directly applicable to his own problems.

For almost every topic discussed in the following pages the data are insufficient. The student who attempts interpretations under these circumstances does so in the face of certainty that some of his conclusions will be rejected. It is, however, pusillanimous to avoid making our best efforts today because they may appear inadequate tomorrow. Indeed, there would be no tomorrow for science if this common attitude were universal. Facts are useless to science unless they are understood. They are to be understood only by theoretical interpretation. The data will never be complete, and their useful, systematic acquisition is dependent upon the interpretation of the incomplete data already in hand. The one merit that is claimed for this study is that it suggests new ways of looking at facts and new sorts of fact to look for.

Tempo and Mode in Evolution

Chapter I: Rates of Evolution

HOW FAST, as a matter of fact, do animals evolve in nature? That is the fundamental observational problem of tempo in evolution. It is the first question that the geneticist asks the paleontologist. Some attempt to answer it is a necessary preliminary for the whole consideration of tempo and mode. Answers can be given, not general answers, but those derived from studies of rates in a few typical cases. These answers are useful, they are suggestive of broader fields for study, and they will serve to supply concrete examples for reference throughout this work. Yet they are not fully satisfactory, because here at the very beginning the study of evolutionary tempo comes up against the gravest sorts of difficulty. There have been many sporadic enquiries, but there has been no systematic accumulation of the needed data. More fundamentally, the concept of rate of evolution is complex and involves many inherent impediments, as will be seen.

Rate of evolution might most desirably be defined as amount of genetic change in a population per year, century, or other unit of absolute time. This definition is, however, unusable in practice. Direct study of genetic change is impossible to the paleontologist. It is, so far, possible only to a limited extent among living animals, and these usually do not change sufficiently in the time available to yield a reliable measure of rate. Even rough estimation has some usefulness, but absolute time cannot yet be accurately measured for paleontological materials, and it is desirable to examine relative as well as absolute rates.

For present purposes, then, rate of evolution is practically defined as amount of morphological change relative to a standard. It is assumed that phenotypic evolution implies genetic change and that rates of evolution as here defined are similar to, although not identical with, rates of genetic modification. This morphological approach to these particular problems is logical, because it is the organism, the phenotype, as an agent in an evolving world with which we are directly concerned. The implicit genetic factors are important less for their own sake than because they are determinants of phenotypic evolution.

The morphological changes studied may be in a single zoological character, giving a unit character rate, or in a number of related characters, a character complex rate, or in whole animals, an organism rate. They may be followed in single lines of descent, a phyletic rate, or averaged over a larger taxonomic unit, a group rate. The scale of comparison or rate division may be absolute time, giving an absolute rate, or an external variate correlated with time, a correlative rate, or an associated change within the group studied, a relative rate. Paleontological data will be adduced to exemplify each of these different kinds of evolutionary rate.

RELATIVE RATES IN GENETICALLY RELATED UNIT CHARACTERS

In any phyletic series various different characters are changing over the same period of time. Paleontologists have long noticed that two such characters may evolve in such a way that the direction and rate of change in one are functions of the change in the other. Various theories of orthogenesis, purposive evolution, and the like have been based on such observations. The methods of analysis of relative growth, largely developed and summarized by Huxley (1932), have cast unexpected light on these phenomena. It has been found that the varying relative sizes of different structures are frequently determined by a constant relationship of their growth rates. This suggests that changes in proportions in evolution may be determined in the same way.[1] Paleontological study along these lines has barely begun, but it already amply demonstrates that this is true in many cases.

It is well known that the sequence from *Hyracotherium* to *Equus* involves increase in gross size and accompanying increase in the length of the muzzle relative to the cranium. Robb (1935) has expressed and studied this phenomenon of "progressive pre-optic predominance" in terms of relative growth. His work shows that the absolute rates of increase of the muzzle length and of the total skull length are different, but that they tend to maintain a constant ratio to each other—that is, that the relative growth of these two parts tends to be constant. The relationship can be approximately expressed by the equation $Y = .25\ X^{1.23}$, in which Y is the preorbital length, X the skull length, and 1.23 the ratio of the rate of increase in Y to that in X. Practically the same equation applies not only (1) to successive stages in a single

[1] Not because of any direct analogy between ontogeny and phylogeny, but because the structure of every adult individual in the evolutionary series is the result of its ontogeny, and ontogeny is hereditary.

phyletic sequence leading to *Equus* but also (2) to contemporaneous equid races of different sizes and (3) to the ontogenetic development of *Equus caballus* (see Fig. 1). It has not been demonstrated in this case, but is a corollary of these demonstrations and is known to be

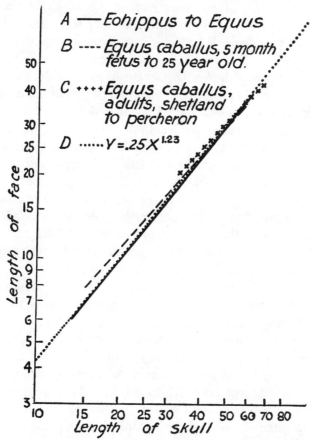

Fig. 1.—Relative rates in horses. Double logarithmic graph of regression of length of face on length of skull, Robb's data: *A*, phylogeny in the line *Hyracotherium-Equus*; *B*, Ontogeny in *Equus caballus*; *C*, races of *Equus caballus*, adults; *D*, heterogony formula nearly fitting all three observed regressions. (Lines *A-C* fitted by eye.)

true in numerous analogous cases, that the same relationship holds true (4) between adults of different sizes belonging to a single race.

These data make it so probable that it amounts almost to a proven fact that the relative sizes of the two variates in question are genetically related or have proportions resulting from a single genetic rate determinant. In the horse skull there was no evolution of this character, which seems to have been constant throughout the whole family for some 45 million years. The striking changes in skull proportions were only the result of the sizes reached by adults of the various species, and evolution in the size of the average adult did occur.

Subsequent similar work by Robb (1936) on the proportions of the digits in the horse foot revealed a still more significant set of phenomena. In three-toed horses the length of either side toe (digits II and IV) is a relatively simple function of the length of the cannon bone (metapodial of III): $Y = 1.5\ X^{.97\ \text{to}\ .98}$, in which Y is the total length of either lateral digit and X the length of the cannon bone of the same foot. This is similar to the previous result for skull proportions, except that the heterogeny is negative and less in degree—indeed, it is not established that the heterogony is significant. The side toes maintained nearly or quite a constant ratio to the cannon bone. The reduction is not surely progressive, as has been universally stated, and the side toes nearly retained the ancestral proportions as long as they were functional at all. But in the one-toed horses the relationship of splints to cannon bone is abruptly different: $Y = .76\ X^{.99\ \text{to}\ 1.00}$.

The constant b (of the formula $Y = bX^k$) is only one-half as large, while k is about the same. Thus, the change from three-toed to one-toed is not simply related to size or to foot elongation, but involves a particular genetic change or mutation, related to this character as a separate entity. In a later paper (1937) Robb shows that this relationship is essentially the same in recent horse ontogeny and between recent horses of different sizes: $Y = .75\ X^{.99}$. Thus unlike skull proportions, digit proportions did evolve, in themselves, but they did so only in one stage of horse evolution and as far as has yet been observed, in one step.

Phleger (1940) and Phleger and Putnam (1942) have made extensive studies on relative growth rates in felids and oreodonts. In these groups they found marked differences in homologous rates in related species and genera, demonstrating evolution and differentiation more extensive than that in the two rates studied by Robb. Phleger's data also hint, although they are insufficient to prove, that different relative growth rates may occur within one population, as if they were the expression of alleles.

In such cases—and they will surely be found to include many of the characters used in the paleontological study of evolution—the impression as to rates of evolution derived from simple observation of morphology may be seriously misleading. Study of the evolution of equid preorbital dominance is a study of a character that did not evolve genetically but only morphologically. In the study of similar proportions in oreodonts, inspection of individuals cannot determine whether genetic evolution has occurred or not unless, as is rarely true, skulls of exactly the same size are available in successive stages. The true rate of evolution in these examples is the rate of change in regression, not that of change of proportion in adults, which is the resultant of regression and of mean adult size.

The criteria for recognition of this situation are that the regression tends to be the same within individuals, between individuals of one population, and also, except as modified by a recognizable mutational advance, between related contemporaneous populations and between ancestral and descendent successive populations. With paleontological materials the intraindividual regression is not available, but an analogous test may be made by comparing the regression of growth stages with that between individuals in a single growth stage and both of these with regressions between distinct contemporaneous and successive populations. It is not true that all relative rates of evolution are determined by a single growth factor governing the development of the two or more characters involved, as will appear from the next example.

RELATIVE AND ABSOLUTE RATES IN GENETICALLY INDEPENDENT CHARACTERS

Another striking progressive character in the evolution of the horse is hypsodonty, increase in height of cheek tooth crowns relative to their horizontal dimensions. The situation here is more complex than in the examples of relative evolutionary rates studied by Robb, Phleger, and Putnam. Both vertical and horizontal dimensions of these teeth are positively correlated with gross size of the animal (and hence with almost all its other linear dimensions) in all three of the possible ways: within populations, between contemporaneous populations, and between successive populations. These tooth characters have no genetically controlled intra-individual variation. Hypsodonty, the relationship between vertical and horizontal dimensions, is positively correlated with size and with most linear dimensions among successive populations, but shows no such correlations among individuals or among contemporaneous populations.[2] The successive intergroup correlation is thus spurious, like so many correlations between temporal sequences. Hypsodonty and size both developed progressively but did so independently. The horses became larger and more hypsodont, but the two characters are separately determined in a genetic sense. Any real relationship was indirect and nongenetic, for instance through natural selection because greater hypsodonty assists the survival of larger animals.

Hypsodonty is one of the most important elements in horse evolution (with size, coronal pattern, and foot structure), but there are few good data on it. For the present illustrative purposes five small samples have been selected from the American Museum collections and the essential data gathered and analyzed. Among many possible measures of hypsodonty, the following index was selected as best adapted to the available material: $100 \times$ (paracone height)/(ectoloph length).

Measurements were made on unworn M^3. The samples have the following identifications and specifications:

Hyracotherium borealis: Lower Eocene, Graybull Formation, Bighorn Basin, Wyoming

Mesohippus bairdi: Middle Oligocene, Lower Brulé Formation, Big Badlands, South Dakota

Merychippus paniensis: Upper Miocene, associated in a "Lower Snake Creek" deposit, Nebraska

Neohipparion occidentale: Late Lower or Early Middle Pliocene, associated in an "Upper Snake Creek" deposit, Nebraska

Hypohippus osborni: Upper Miocene, associated in a "Lower Snake Creek" deposit, Nebraska

Hyracotherium-Mesohippus-Merychippus-Neohipparion and *Hyracotherium-Mesohippus-Hypohippus* represent approximate genetic phyla. *Neohipparion* and *Hypohippus* are thus typical of divergent phyla of common ancestry (slightly beyond the *Mesohippus* stage).

[2] This statement agrees with, but is not proven by, calculated correlation coefficients. Hypsodonty can only be measured as a ratio, and the statistical correlation of a ratio with one of its elements or with a variate correlated with the latter is frequently spurious. Nevertheless, the stated independence is an evident and, I believe, incontrovertible biological fact.

Rates of Evolution

Some of the pertinent statistics[3] are given in Table 1:

TABLE 1
MEASUREMENTS ON M³ OF FIVE SAMPLES OF FOSSIL HORSES

A. PARACONE HEIGHT IN MILLIMETERS

	N[a]	O.R.	S.R.(S.D.)	M	σ
Hyracotherium borealis	11	4.2- 5.1	1.9	4.67 ± 0.09	0.29 ± 0.06
Mesohippus bairdi	14	7.8- 9.4	2.6	8.36 ± 0.11	0.40 ± 0.08
Merychippus paniensis	13	29.6-37.6	13.0	34.08 ± 0.56	2.01 ± 0.39
Neohipparion occidentale	5	49-55	15.6	52.40 ± 1.08	2.41 ± 0.76
Hypohippus osborni	4	16.7-22.4	17.0	18.75 ± 1.31	2.62 ± 0.93

B. ECTOLOPH LENGTH IN MILLIMETERS

Hyracotherium borealis	11	7.6- 8.9	3.0	8.21 ± 0.14	0.46 ± 0.10
Mesohippus bairdi	14	11.0-13.0	3.6	11.89 ± 0.15	0.55 ± 0.10
Merychippus paniensis	13	17.7-21.7	6.9	19.96 ± 0.29	1.06 ± 0.21
Neohipparion occidentale	5	19-22	7.1	20.80 ± 0.49	1.10 ± 0.35
Hypohippus osborni	4	19.4-26.4	20.5	22.03 ± 1.59	3.17 ± 1.12

C. 100 x A/B

Hyracotherium borealis	11	54-60	11.7	57.0 ± 0.5	1.8 ± 0.4
Mesohippus bairdi	14	64-77	20.7	70.4 ± 0.9	3.2 ± 0.6
Merychippus paniensis	13	155-184	48.0	170.7 ± 2.1	7.4 ± 1.5
Neohipparion occidentale	5	241-262	61.6	252.2 ± 4.3	9.5 ± 3.0
Hypohippus osborni	4	84-88	11.0	85.5 ± 0.9	1.7 ± 0.6

[a] N, size of sample; O.R., observed range (by extreme measurements); S.R.(S.D.), standard range from standard deviation (by span); M, mean; σ, standard deviation. The last two with standard errors.

The same data are graphically shown in Fig. 2, set up to represent the hypothesis that the (geometric) growth rate of ectoloph length was constant (by placing these values in a straight line on semilog coordinates). The assumption that gross size increase (with which ectoloph length is closely correlated) was approximately constant in rate has been made for this and other so-called orthogenetic series. The diagram shows, in fact, that the hypothesis is false, for if it were true the horizontal distances between species would be proportionate to the geologic ages, whereas *Hypohippus osborni* comes out much too far to the right, the distance from *Neohipparion occidentale* to *Merychippus paniensis* is surely too small, and that from the latter species to *Mesohippus bairdi* is probably too large, relative to the *Hyracotherium-Mesohippus* distance. The more important conclusion from the diagram is, however, that the rate of evolution of height behaved in a very different way from that of ectoloph length. Plotting these as for

[3] Here and elsewhere, it is assumed that the reader is familiar with elementary statistics. For an introduction to this subject, see Simpson & Roe (1939).

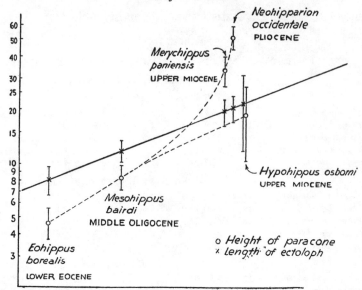

FIG. 2.—Relationships of structural changes in two characters of five genera of Equidae. Height of unworn paracone and length of ectoloph of M^3, data in text. Ordinate scale logarithmic, no abscissal scale; arranged in hypothesis of rectilinear mean increase in ectoloph length. Circles and crosses are means of samples. Vertical lines are standard ranges, statistical estimates of variation in a population of 1,000 individuals.

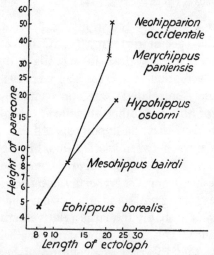

FIG. 3.—Relative rates of evolution in equid molars. Same data as for Fig. 2 (means of samples only) with paracone height plotted against ectoloph length on double logarithmic co-ordinates.

relative growth (Fig. 3), it is seen that the rate for paracone height was higher than for ectoloph length throughout, that the ratio of these rates was approximately constant in the *Hyracotherium-Hypohippus* line, but that the rate for paracone height showed differential acceleration in the *Mesohippus-Neohipparion* line.

In Fig. 4 an attempt is made to show the true temporal trends in these rates. The plot is semilog, with time on the arithmetic scale. The

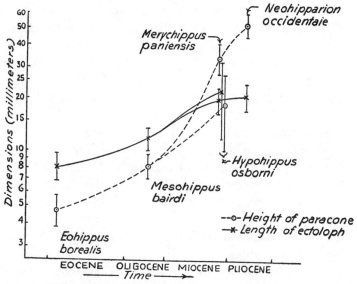

FIG. 4.—Phylogeny of five genera of Equidae and evolution of two related tooth characters. Data as for Fig. 2. Ordinate scale, logarithmic, absolute dimensions. Abscissal scale, arithmetic, approximation of absolute lapse of time (probable relative lengths of epochs represented in arithmetic proportion). Slopes of lines are proportionate to rates of evolution of the unit characters.

precise absolute or relative lengths of the Tertiary epochs are unknown, but the relative lengths were probably more or less as shown. It is seen that neither rate was constant in the two phyletic series involved. Both show acceleration toward the end of the Oligocene. The acceleration in increase of paracone height was greater and continued until about the Middle Miocene. Both show deceleration in the Late Miocene; this is more pronounced in ectoloph length, so that increase of this dimension almost ceases in the Pliocene. The rate is more nearly constant for ectoloph length than for paracone height. For ectoloph length the rates in the two phyla diverge very little—the data are insufficient to prove that the small apparent divergence is not due to chance. For paracone height, however, the rates in the two phyla diverge markedly and significantly. In the Miocene this dimension increased much more rapidly in the line leading to *Neohipparion* than in that leading to *Hypohippus*. The result is that *Hypohippus*, although decidedly more hypsodont than the common ancestry of the two lines, is much less hypsodont than the contemporary *Merychippus*.

Hypsodonty is a single character from a physiological point of view and it is unquestionably a unit in its reaction with the pressure of natural selection; but it here appears as a resultant of two other and simpler characters evolving with considerable, not complete, independence from each other. The probable reasons for the slow advance of hypsodonty throughout and its marked acceleration in one line during the Miocene will be discussed in a later section.

Study of the other essential differences between *Merychippus* and *Hypohippus* shows that most of them arose because of differences in rates of evolution, both the rates and the differences between them being distinctive for each character. This is a widespread evolutionary phenomenon in diverging phyla of abundant animals evolving at moderate rates.

The Equidae are thus found to illustrate four basic theorems concerning rates of evolution:

1. The rate of evolution of one character may be a function of another character and not genetically separable even though the rates are not equal.

2. The rate of evolution of any character or combination of characters may change markedly at any time in phyletic evolution, even though the direction of evolution remain the same.

3. The rates of evolution of two or more characters within a single phylum may change independently.

4. Two phyla of common ancestry may become differentiated by differences in rates of evolution of different characters, without any marked qualitative differences or differences in direction of evolution.

CORRELATIVE RATES

A correlative rate can be determined when the values of a morphological variate are correlated with some variate external to the animals and the latter variate is, in turn, correlated with time. The usual external variate is thickness of strata, and rates relative to such thicknesses have frequently been used, especially for marine invertebrates. That the stratigraphic succession is temporal, when correctly determined, is a proven fact, but the validity of the method also demands a constant relationship between not only succession but also thickness of strata and time, that the rate of deposition shall have been approximately constant. This is difficult to prove and frequently is not true, so that the usefulness of the method is sharply limited. Moreover, the usual proof of sedimentary continuity is faunal continuity, and then the determination of rates of evolution relative to sedimentary units is invalid. Nevertheless, the method can give some information about fluctuations in rates of evolution, and it is a valid way of comparing the rates of different characters or of different phyla during the same span of time (identical sequences of strata).

Demonstration that a character has significantly evolved in a given series of strata may be made, among other ways, by calculation of a correlation coefficient or by analysis of variance into intersample and intrasample, the samples being successive. These figures do not measure any rate of evolution, however, and of several possible measures of rate probably the most useful is the regression coefficient in cases of nearly rectilinear regression against thickness of strata.

One of the most extensive and completely analyzed sets of data from which correlative rates can be obtained was given by Brinkmann (1929) for the ammonite genus *Kosmoceras* through about 13 meters of predominantly fine-grained sediments near Peterborough, England. Table 2 presents typical data, rearranged and somewhat revaluated, from statistics calculated by Brinkmann.

Only for the groups of strata in which the correlation coefficient is significant do the data demonstrate that any change in the character took place (i.e., that either the correlation coefficient or the regression coefficient is likely to exceed zero). The value of such subdivided data is that they reveal that rates have changed, not simply that evolution did occur or what its average value is over a long span. In dealing with a single character, comparisons are valid only if no hiatus is present within any one class of the stratigraphic sequence. This is not apparent from the table itself, but other information in Brinkmann's paper shows that there is no significant hiatus within this table.

TABLE 2

CORRELATION AND REGRESSION WITH THICKNESS OF STRATA FOR TWO CHARACTERS OF *Kosmoceras* (*Zugokosmoceras*)

Character		Strata Distance in cm. from bottom of section.	N^a	r	b
I Terminal diameter	b. c. d. e. f.	26-28 29-39 40-50 56-78 79-134	23 32 25 19 32	.32 .34 .24 .45^b .47^c	3.0 ± 1.8 0.55 ± 0.26 0.58 ± 0.46 1.6 ± 0.6 0.22 ± 0.06
II Diameter at disap- pearance of outer nodes	a. b. c. d. e. f.	7-20 26-28 29-39 40-50 56-78 79-134	35 67 80 74 147 96	.08 .07 .34^c .14 .52^c .23^b	0.10 ± 0.22 0.40 ± 0.70 0.55 ± 0.16 0.33 ± 0.28 0.96 ± 0.11 0.11 ± 0.05

^{a} N, size of sample; r, correlation coefficient; b, regression coefficient, change of character in millimeters per 1 cm. of strata. ^{b} Significant. ^{c} Highly significant.

For the terminal diameter (character I), the table shows that change was certainly occurring while strata groups *e* and *f* were laid down and that the regression was considerably faster in *e* than in *f*. It does not show that any change occurred in *a–d*, inclusive, or whether the regression was then faster or slower than in *e* and *f*. For the diameter at disappearance of outer nodes (character II), change is shown to have occurred in *c*, *e*, and *f*, and regression was faster in *e* than in *f*. In *c* the rate was probably intermediate, but it is not certain that it was either slower than in *e* or faster than in *f*. The rate was probably slower in *a* than in *e*. Both characters I and II were evolving at about the same rate and accelerated and decelerated together.

Since these rates are relative to thicknesses of strata, the demonstrated changes in them may mean either that rate of evolution changed or that rate of sedimentation changed. If the two characters were not correlated with each other, the tendency for rates to vary together would suggest (but not prove) that the variation was mainly in sedimentation. In fact, they are highly correlated with each other ($r = +.66 \pm .15$ for beds 65–70, where the greatest change in regression occurs, and $r = +.85 \pm .02$ for the whole sequence). Other data show that there are, for some characters at least, real changes in rate of evolution. For instance, in the subgenus *Kosmoceras* (*Anakosmoceras*) regression of the "bundling index" (*Bündelungsziffer*) is negative in strata 1080–1093 and positive thereafter, a change that cannot be caused by rate of sedimentation. In other cases a fairly high positive regression is followed or preceded by one so nearly zero over so long a sequence of strata that explanation by rapidity of sedimentation is incredible. For instance, after strata 26–134 of the table, in which it shows well-marked regression, terminal diameter in *Kosmoceras* (*Zugokosmoceras*) shows no significant regression from 136 to 380, a sequence of strata more than 2½ times as thick.[4]

The ammonites form one of the groups in which great regularity of evolution, as to both direction and rate, have been claimed. Brinkmann's materials afford exceptionally good conditions for the demonstration of this regularity, if it exists, and his data are remarkably complete and objective. Although they cannot rigidly prove irregularity, because alternative explanations cannot be completely ruled out, they strongly suggest it. In any event, they fail to confirm the usual conclusion, based on fewer observations and far more subjective methods of inference.

ORGANISM RATES

Direct determination of rate of evolution for whole organisms, as opposed to selected characters of organisms, would be of the greatest value for the study of evolution. Matthew wrote, nearly a generation ago (1914), "to select a few of the great number of structural differences for measurement would be almost certainly misleading; to average them all would entail many thousands of measurements for each species or genus compared." On the basis suggested in this quotation, the problem would be immediately soluble in theory and probably also in practice, because the taking of thousands of measurements is not an insuperable difficulty and there are now methods for reducing them to coherent and easily manipulated form. As the situation is now understood, the most serious difficulties are (1) selecting unit characters for measurement, (2) reducing them all to metrical form in comparable units, and (3) weighting them in order to obtain a valid

---

[4] This character, like several others, also shows reversal of trend; but, since it occurs during a hiatus, this may have been caused by local extinction and repeopling by a less progressive stock from elsewhere.

general average. Many paleontologists and zoologists have proceeded as if everything that can be measured or observed as a unit were a unit character. As preceding examples show, some characters, in this sense, are highly correlated with each other, some slightly correlated, and some independent. In study of the organism as a whole, some account must be taken of these correlations in determining the number and nature of unit characters. Some criterion of classification is needed. For instance, hypsodonty is an important unit character on the criterion of selection value, but morphologically it is the resultant of two other characters which are correlated in one way and uncorrelated in another, and genetically it is undoubtedly controlled by at least two and probably many genes that simultaneously control other, quite distinct phenotypic characters. Characters such as tooth pattern, color, cranial angulation, number of vertebrae, and length of limbs cannot be measured in the same units and must be reduced to a common relative form before averaging is possible. Some essential characters, such as tooth pattern, are difficult to measure in any unit. Finally, an unweighted average might be very misleading because some morphological characters are more crucial, more constant, more independent, more strictly hereditary, and so forth, than others.

It cannot be said that the problem is quite insoluble, but certainly it is so complex and requires so much knowledge not now at hand that no solution is in sight at present. It is still true, as when Matthew wrote, that subjective judgment of the total difference between organisms is (if made by an able and experienced observer) more reliable than any objective measurement yet devised. Obviously, only rough approximations can be made in this way; but an approximation, recognized as such, is more useful than a seemingly exact but really spurious average or no measure at all. Insofar as it seeks to divide phyla into generic and specific stages, representing roughly equivalent amounts of total morphological change, the taxonomic system is a rich source of such data. The assumption that two successive or related genera do cover equivalent amounts of evolution is obviously very uncertain in any one instance. It becomes more reliable and useful when the taxonomists involved are of equal and great skill, when one student with extensive first-hand knowledge has revised all the genera, and especially when a large number of genera based on more or less comparable criteria can be averaged. For such purposes genera are the most useful units. For paleontological materials, at least, they are more clearly defined and more nearly comparable than any others at present, and they fill the further requirements that they are intended to be essentially monophyletic in origin, to have an extension in time, and to be horizontally divided from preceding and following units of the same rank.

If all genera were strictly comparable, organism rates of evolution would be proportional to the reciprocals of the durations of the genera in question. For a sequence of successive genera, a more reliable value would be obtained by dividing number of genera by total duration. Thus in the line *Hyracotherium-Equus* (but omitting *Equus* because its span is incomplete and indeterminate) there are eight successive genera according to good modern classifications (e.g., Stirton 1940). The time covered is about 45,000,000 years, and the average rate can therefore be expressed as .18 genera per million years, or reciprocally as 5.6 million years per genus.

Strictly comparable averages can only be obtained for genera that arise at known times from known ancestors and that disappear not by extinction, but by evolution into other genera. The number of such genera now known in any one group is small. Some comparisons are nevertheless possible as suggested by the data in Table 3.

TABLE 3

RATES OF EVOLUTION IN HORSES, CHALICOTHERES, AND AMMONITES IN TERMS OF NUMBER OF GENERA PER MILLION YEARS

Group or Line	Number of Genera[a]	Average Genera (in one line); per Million Years	Phylogeny and Classification Used as Basis
Hyracotherium-Equus	8	.18	Stirton 1940
Chalicotheriidae	5	.17	Colbert 1935
Triassic and earlier ammonites	8	.05	Swinnerton 1923 and others

[a] With approximately known time of origin and time of transformation into another genus.

With due allowance for all the uncertainties involved, it is safe to conclude that the rate of evolution in chalicotheres was about the same as in horses and that it was faster in both groups of perissodactyls than in the early ammonites.

Analogous estimates of changes in rate within a single line are less useful and require weighting. Matthew (1914) tried this with horse genera, although his purpose was the reverse of the present attempt: he postulated a uniform rate of evolution and attempted to estimate lapses of time by relative amounts of evolution. Matthew's figures were severely criticized by Abel (1929) first because of their subjective and approximate value, and, second, because of some differences of opinion, e.g., that the step *Parahippus-Merychippus* should have been relatively longer, *Eohippus-Orohippus* and *Orohippus-Epihippus* relatively shorter. Despite the very rough nature of such approximations, which was freely admitted by Matthew as it is here, they do have considerable interest and an attempt to revise the estimates on more recent data may have some value.

TABLE 4

ESTIMATES OF CHANGES IN RATES OF EVOLUTION IN GENERA OF HORSES

Genus	A^a	B	C	D
Equus	10	7	6	1
Pliohippus[b]	10	11	10	1
Merychippus	15	18	6	3
Parahippus	5	5	4	1
Miohippus	5	5	4½	1
Mesohippus	15	16	4½	4
Epihippus	10	9	5	2
Orohippus	10	9	5	2
Hyracotherium[c]				

[a] A, Matthew's weighting, on a basis of an average weight of 10 per genus. In each case the weight is understood to be for the approximate total advance to the midpoint of this genus from the midpoint of that preceding. B, similar weights adjusted to subsequent criticism and discovery. C, estimates of approximate time involved, in millions of years. D, rates obtained by dividing the adjusted weights (B) by the approximate time (C).
[b] *Hipparion* in Matthew. *Pliohippus* is now known to be nearer the direct line and its evolutionary stage is roughly comparable.
[c] *Eohippus* in Matthew. *Eohippus* and *Hyracotherium* are now believed to be synonymous.

On the whole, the rates thus obtained are reasonable relative values, at least to the point showing more rapid average evolution in the middle to late Eocene than subsequently and acceleration in the late Eocene to early Oligocene and early to middle Miocene.

Reverting to Matthew's original purpose, Table 5 gives estimates of the lengths of the Tertiary epochs, except Paleocene, postulating a total of 45,000,000 years.

It seems almost certain that the Miocene was considerably longer

TABLE 5
ESTIMATES OF DURATIONS OF TERTIARY EPOCHS
In millions of years

	A*	B	C
Pliocene	10	8	9
Miocene	11	13	13
Oligocene	7½	8	7
Eocene	16	16	15

* A, based on horse genera, Matthew's weighting, on his hypothesis of uniform evolution. B, same, revised weighting (B of preceding title). C, independent estimates based on a balance of all available evidence (sedimentation, radioactivity, general faunal change, etc.).

than the Pliocene and that the estimates based on Matthew's weighting and hypothesis are defective to that extent (but this may be partly due to a different placing of the boundary, on which Matthew frequently changed his usage). The general agreement of all three estimates, except for this one point, is striking. Agreement between the adjusted weights and the independent estimates suggests that the average rates of horse evolution in each epoch did not differ greatly despite acceleration and deceleration during periods that were less than an epoch or that overlapped epochs.

A number of rather isolated approximations of organism rates of evolution have been accumulated from the study of degree of taxonomic differentiation of groups that moved away from a parent stock at approximately known times. An example fairly typical both in its positive and its negative aspects has recently been provided by Doutt (1942). A stock of the normally marine seal *Phoca vitulina* became isolated in a fresh-water lake in northern Canada at a time variously estimated at 3,000 to 8,000 years ago. The average length of generations in this species is unknown, but it is probably five to ten years. These lake seals have thus been isolated for from 300 to 1,600 generations, or, as a mean estimate, for approximately 1,000 generations. They are in some respects outside the known range of variation in the survivors of the parent marine stock, and their general morphological differentiation is given subspecific rank. Scattered data on rodents similarly isolated (for instance, on islands) give roughly comparable results and suggest that approximately subspecific morphological differentiation may occur in even less than 300 generations, the lowest figure warranted for these seals.

If such observations can be multiplied and made more precise, they may establish a maximum organism rate for various groups. The conditions that provide such data are, however, unusual historic accidents, and considerable hesitation is proper in considering the bearing of such cases on rates of evolution under more normal circumstances.

Something may also be learned of organism rates from the study of events such as the entrance of North American animals into South America in the Pliocene and the Pleistocene. Many mammals in stocks that entered the southern continent between one and two million years ago have developed endemic genera there, but none has developed endemic families. The conclusion is therefore justified that this length of time has sufficed for generic, but not much higher, differentiation under these circumstances and for these groups, e.g., cricetine rodents, procyonid carnivores, and deer. It is curious that this example gives no clear evidence for the more rapid evolution of animals with shorter generations. For instance, the small rodents apparently did not evolve more rapidly than ungulates with generations several times as long.[5]

Much of the evidence of this sort is so vague and unsatisfactory that its interpretation is almost entirely subjective, and the same facts may be used to reach diametrically opposed conclusions. Thus the great endemism in faunas of isolated Pacific islands has been cited as exemplifying slow evolution on very old islands and also as proof that evolution has there occurred very rapidly (e.g., Zimmerman 1942).

GROUP RATES AND SURVIVORSHIP

Estimates of the average duration of genera within a phylum, as above, are one sort of group rates of evolution and on the whole probably the most satisfactory when obtainable. They are, however, greatly limited by the small number of genera for which both ancestral and descendent genera are surely recognized. Use of all the genera of a larger taxonomic group introduces other sources of error: (1) most genera certainly have a fossil record shorter than their real duration; (2) the numerous genera that disappeared by extinction cannot, on the average, have undergone evolutionary changes comparable to those of genera that disappeared by transformation; and (3) the fossil record of more slowly evolving genera probably is in general more

[5] The rodents have a greater number of endemic genera, but the genera are no more distinct morphologically.

complete than that of rapidly evolving genera in the same group. The first two sources of error will tend to make estimates of rate too high, and the last to make them too low. The extent of compensation by these opposite tendencies cannot be determined. The errors are, however, systematic and more or less independent of the particular nature of the fossils in question. Thus, they deprive the rate estimates of absolute accuracy, but do not necessarily invalidate the estimates as relative rates in the comparison of different groups. With all their shortcomings, such data do prove to have considerable value and to reveal facts of great importance both for tempo and for mode of evolution, as will be shown.

To explore and illustrate the possibilities, two very different groups

TABLE 6

DISTRIBUTION OF GENERA OF PELECYPODA

Figures are numbers of known genera

		Last Known Appearance									Totals, First Known Appearances
		Ordovician	Silurian	Devonian	Carboniferous	Permian	Triassic	Jurassic	Cretaceous	Tertiary	Recent	
	Ordovician	13	8	6	1	0	4	0	0	0	1	33
First Known Appearance	Silurian	..	17	13	0	1	2	1	0	0	4	38
	Devonian	..	..	30	10	4	5	1	0	0	4	54
	Carboniferous	..	..	..	16	3	0	0	1	0	3	23
	Permian	..	..	..	..	3	1	0	1	0	3	8
	Triassic	..	..	..	..	..	30	6	12	0	20	68
	Jurassic	..	..	..	..	..	..	25	12	3	16	56
	Cretaceous	..	..	..	..	..	..	..	37	0	24	61
	Tertiary	..	..	..	..	..	..	..	..	18	64	82
	Totals, last known appearances	13	25	49	27	11	42	33	63	21	139	423

TABLE 7
DISTRIBUTION OF GENERA OF CARNIVORA (EXCEPT PINNIPEDIA)
Figures are numbers of genera

			Last Known Appearance															
			Paleocene			Eocene			Oligocene			Miocene			Pliocene			Pleisto-cene	Recent	Total First Known Appearances
			L	M	U	L	M	U	L	M	U	L	M	U	L	M	U			
First Known Appearance	Paleocene	L	5	0	0	1	..	..	..	..	..	..	..	..	..	..	..	..	..	6
		M	..	12	0	1	..	..	..	..	..	..	..	..	..	..	..	..	..	13
		U	..	..	5	3	..	..	..	..	..	..	..	..	..	..	..	..	..	8
	Eocene	L	..	..	..	8	5	3	..	..	..	..	..	..	..	..	..	..	..	16
		M	..	..	..	..	7	4	..	..	..	..	..	..	..	..	..	..	..	11
		U	..	..	..	..	..	13	6	1	0	0	2	..	..	..	..	..	..	22

																			Total first known appearances
First Known Appearance	Oligocene	L	·	·	·	·	·	·	9	1	4	5	2	0	1	·	·	·	·	22
		M	·	·	·	·	·	·	·	3	·	·	·	·	·	·	·	·	·	3
		U	·	·	·	·	·	·	·	·	0	3	1	·	·	·	·	·	·	4
	Miocene	L	·	·	·	·	·	·	·	·	·	20	0	3	2	·	·	·	·	25
		M	·	·	·	·	·	·	·	·	·	·	9	3	7	1	1	·	·	22
		U	·	·	·	·	·	·	·	·	·	·	·	8	8	1	0	2	·	19
	Pliocene	L	·	·	·	·	·	·	·	·	·	·	·	·	23	5	4	1	7	40
		M	·	·	·	·	·	·	·	·	·	·	·	·	·	4	2	0	2	8
		U	·	·	·	·	·	·	·	·	·	·	·	·	·	·	4	2	3	9
	Pleistocene		·	·	·	·	·	·	·	·	·	·	·	·	·	·	·	15	19	34
Total last known appearances			5	12	5	13	12	20	15	5	4	28	14	14	41	11	11	19	33	262

may be taken: pelecypod molluscs and carnivorous placental mammals (excluding the pinnipeds, for which the record is wholly inadequate). The raw data consist of the geological distributions of all the known genera in each group. These data are summarized in the accompanying tables, in which the numbers of genera running through any given sequence of the geological time scale are entered. Data on pelecypods were gathered chiefly from the latest editions of the standard Zittel, *Grundzüge,* in German, English, and Russian revisions. These genera are broadly drawn and not exhaustively listed, but the data are sufficiently good for present purposes and prove to be adequately enlightening. Data on carnivore genera are more complete and accurate, having been taken from my unpublished classification of mammals, which in turn is based upon almost all the literature of the subject. In both cases the many recent genera that are not known as fossils are omitted. Among the pelecypods, genera unknown before the Pleistocene are also omitted, and the Pleistocene data for the carnivores are less complete than for the Tertiary and not entirely comparable.

Because the lengths of the various periods and epochs differ greatly, these tables do not directly yield estimates of rates of evolution. The geological ages must be translated into terms of relative or absolute time, which introduces another source of error, since this translation cannot as yet be exact. The available estimates of geological chronology are, however, good enough to warrant their use. For estimates of relative durations see, e.g., Schuchert and Dunbar (1933), and for the present status of absolute age measurements by radioactivity see Goodman and Evans (1941).

One interesting method of presentation and analysis of these data after statement in terms of years of duration is by modified survivorship curves (as explained, for instance, in their customary form in Pearl 1940). One method of construction of such curves adapted for the present use is shown by the solid lines in Fig. 5. Here only genera now extinct are counted, and the plotted points represent the percentage of all these genera with a given known duration equal to or higher than the various stated numbers of years. The actual curves approximating these points have been roughly sketched in by eye. Although similar in form, the curves for pelecypods and carnivores differ greatly in extent, the mean survivorship for a genus of Pelecypoda being 78 million years and for a genus of Carnivora only 6½ million years.

The data undoubtedly exaggerate the difference, for various reasons, but it is safe to say that carnivores have evolved, on an average, some ten times as fast as pelecypods.

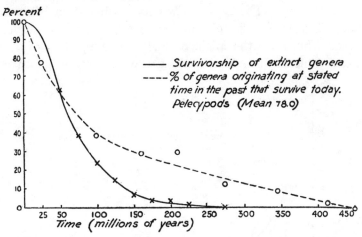

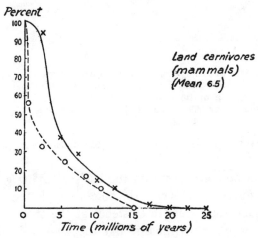

Fig. 5.—Survivorship curves for genera of pelecypods and of land carnivores. Continuous lines, survivorship in genera with completed span (extinct); broken lines, survivorship on basis of ages of genera now living (and known as fossils). Crosses and circles are calculated values to which curves are roughly fitted. Arithmetic co-ordinates; time scale absolute.

The similarity of the curves is more clearly shown and the existing differences are revealed by replacing the absolute time dimension by deviations expressed in percentage of average survivorship, as in Fig. 6

(Pearl 1940). For further comparison an analogous curve for survivorship in a population of mutant *Drosophila* is also given and is found to be closely similar to the generic survivorship curves, especially that for pelecypods. The *Drosophila* curve is based on life spans of individual flies and so is only analogous to the generic curves, not homologous; but the latter might be said to give a picture of a sort of evolutionary metabolism in the two groups concerned, much as the

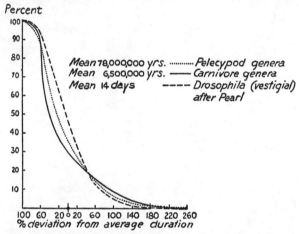

FIG. 6.—Survivorship in pelecypod genera, land carnivore genera, and *Drosophila* individuals. Reduced to comparable form with mean survivorship of three groups coinciding on scale and time represented by percentage of deviation from this point.

Drosophila curve portrays a sort of vital metabolism in the corresponding population.[6]

Analogous curves can be constructed on the basis of living genera that are also known as fossils. In this case the points are plotted as the percentages of such genera that were in existence at the stated times in the past. In other words these points and the sketched curves (broken lines in Fig. 5)[7] represent the length of time living genera

[6] This does not involve and is not intended to endorse the interpretive use of the supposed analogy between the individual life cycle and the so-called racial life cycle, an analogy that seems to me misleading and commonly misused. Similarity in shape of the curves has no bearing on similarity of the vital or evolutionary processes producing them, as is amusingly illustrated by Pearl (1940, Fig. 84), who shows that survivorship of automobiles follows almost the same pattern as that of cockroaches. A similar fallacy underlies Willis's evolutionary interpretation of his "hollow curves" (1940).

[7] The fit is not as smooth as in the other curves, probably because of the smaller number of genera involved.

have now survived. If the recent faunas were random samples of populations similar, as regards generic survivorship, to the extinct genera of the same groups, curves constructed in this way should approximately coincide with those constructed in the previous way. Obviously they do not coincide, and the differences are significant for the study of evolution. These differences are perhaps shown still more clearly in Table 8, in which the expectation of survival is based on the generic survivorship curves for extinct genera.

TABLE 8

EXPECTED AND ACTUAL GENERIC SURVIVORSHIP IN PELECYPODA AND CARNIVORA

Time	Genera Appearing	Percentage of Approximate Expectation of Survival to Recent	Expected Survivals	Actual Survivals
CARNIVORA				
Early Miocene	..	0	0	0
Middle Miocene	22	2	0	0
Late Miocene	19	15	3	2
Early Pliocene	40	23	9	7
Middle Pliocene	8	37	3	2
Late Pliocene	9	90	8	3
Pleistocene	34	98	33	19
PELECYPODA				
Ordovician	33	0	0	1
Silurian	38	0	0	4
Devonian	54	0	0	4
Carboniferous	23	0	0	3
Permian	8	2	0	3
Triassic	68	3	2	20
Jurassic	56	6	3	16
Cretaceous	61	24	15	24
Tertiary	82	88	68	64

Among the carnivores, survival to Recent agrees sufficiently with expectation for genera that appeared before late Pliocene, but it is much lower than expectation for late Pliocene and Pleistocene genera. The discrepancy was largely, perhaps wholly, caused by the unusually high mortality of the Pleistocene. Among recent pelecypods, on the other hand, survival from the Tertiary agrees well enough with expectation, but survival from all previous periods back to the Ordovician inclusive, is greater than expectation. This means that the living pelecypod fauna, far from having experienced increased mortality, as have the carnivores, includes a large number of very slowly evolving genera and that these slowly evolving lines are less likely to become extinct than are other pelecypods—a striking point of unusual importance, to be discussed in a later section of this study.

Data of this sort also have bearing on differences of rates of evolution within the same group at different times. Although they do not directly measure rates of evolution, it seems probable, a priori, that the number of genera at any one time, the number of genera appearing per million years (rate of origin), and the number disappearing per million years (rate of disappearance) would all be positively correlated with rate of evolution and that the average age of genera at any one time would be negatively correlated with rate of evolution. In Table 9 these figures are given for the Carnivora.

The figures for Lower Paleocene, Pleistocene, and Recent are not exactly comparable with the others.[8] Disregarding these three times,

TABLE 9

SURVIVAL DATA FOR THE CARNIVORA

Time		a. No. of Genera Existing	b. Rate of Origin	c. Rate of Disappearance	d. Average Age of Existing Genera
Paleocene	E[a]	6	1	1	2.5
	M	14	3	2	2.7
	L	10	2	1	3.5
Eocene	E	21	3	3	3.8
	M	19	2	2	3.6
	L	29	4	4	3.6
Oligocene	E	31	8	6	2.2
	M	19	1	2	3.1
	L	18	1	1	4.4
Miocene	E	39	6	6	4.0
	M	33	5	3	4.3
	L	38	5	3	3.5
Pliocene	E	64	13	14	3.8
	M	31	3	4	3.7
	L	29	3	4	4.9
Pleistocene		52	(34)	(19)	(2.7)
Recent[b]		33	..	..	3.4

[a] E = early; M = middle; L = late.
[b] Counting only genera identified in early Pleistocene or earlier.

[8] Carnivores first appeared in the Lower Paleocene and then necessarily had all four figures unusually low, regardless of rate of evolution. The Pleistocene data are not quite

Rates of Evolution

TABLE 10

STATISTICAL SUMMARY OF REPRESENTATIVE DATA ON CARNIVORE EVOLUTION[a]

	R	M	σ
Number of genera	10-64	28.2 ± 3.5	13.1 ± 2.5
Origination rate	1-13	4.2 ± 0.8	3.1 ± 0.6
Disappearance rate	1-14	3.9 ± 0.8	3.2 ± 0.6
Average age	2.2-4.9	3.65 ± 0.18	0.66 ± 0.12

[a] That is, data by stages from Middle Paleocene to Late Pliocene, N = 14 stages in each case.

columns a, b, and c of Table 9 are all highly correlated with each other, as was expected ($r_{ab} = .88$; $r_{ac} = .91$; $r_{bc} = .94$). On the reasonable hypothesis that they are also correlated with the group rate of evolution, the rate of evolution of the carnivores as a whole had three peaks during the Tertiary, in (or just before) early Oligocene, early Miocene, and, especially, early Pliocene; at these times all three of the pertinent figures (columns a, b, c of the Table 9) were simultaneously above their respective means.

Contrary to expectation, column d of the table is not significantly correlated with any of the other three columns. Statistically its fluctuations could be purely random. Biologically it is certainly a result of the factors entered in columns a, b, and c; but the relation is too complex and for each item in column d extends backward throughout too many different items in the other columns for simple analysis of the relationship.[9] The average age of existing genera is an indirect, approximate measure of average group rate of evolution over an indefinite period prior to the time of reference, but changes in this measure do not, in this particular instance, reliably indicate changes in rate of evolution within the group. Averaging over longer periods of time may give some information on this point, but here the data on average age do not mean much more than the single figure for mean survivorship previously obtained. Average age should tend to be approximately one-half the mean survivorship for the same statistical population.

---

complete, and the Pleistocene was much shorter than any previous epoch and otherwise exceptional. The Recent is practically a point in time, and so its data cannot be compared with figures based on duration.

[9] I have made numerous and lengthy attempts to demonstrate the relationship and also to formulate some comparable and valid measure that would show the expected correlation, but the results were negative and therefore are not included.

Chapter II: Determinants of Evolution

AMONG THE MOST IMPORTANT factors that may or do influence both the rate and the pattern of evolution are variability, rate of mutation, character of mutations, length of generations, size of populations, and natural selection. Under "natural selection" may be included numerous different elements determining the intensity and direction of selection, such as environmental change, degree and sort of previous adaptation, presence and nature of competing populations, and so forth. Some other evolutionary determinants, although doubtless important at times, are not specially discussed at this point, because they are believed to be secondary or of less general importance from the point of view of tempo and mode.

VARIABILITY

The endless discussion of the relationship of variability and evolution has been greatly obscured by confusion of concepts and the use of ill-defined terms. Without special definition, "variability" can hardly be taken to mean more than the obvious capacity of individual animals to differ from one another. It is proper, but not consistently customary, to distinguish this capacity, variability, from its expression, variation. Evolutionists have abundantly emphasized that variation may be nonhereditary or hereditary and that the latter fraction is the more important for evolution and is usually meant when a modern evolutionist speaks of variation.[1] Even here the term "hereditary" is equivocal, and it is better to speak of genetic and nongenetic variation, according to whether the differences do or do not represent variation in inheritance

---

[1] Recently it is beginning to appear that the distinction is not, after all, so absolute or so essential. Attention was long centered on whether an acquired character can be inherited as such, but, now that this theory has been as nearly disproved as possible, it is more clearly realized that no morphological character is inherited as such. What is inherited is a complex of potentialities for development, and the ultimate morphological expression of the same hereditary factors may differ markedly. All these variations are "acquired," and all are equally, in this sense, "inherited," not merely the one we consider typical. Variations caused by environment are highly pertinent to the course of evolution. In speaking of them as nonhereditary the distinction now is simply that the particular expression, among several possible expressions, developed in a given individual does not affect the potentialities passed on to its descendants. It certainly affects the fate of the individual and therefore helps to determine whether that individual will have descendants.

as well as in its individual expression. It has now been sufficiently emphasized that genotypic variation and phenotypic variation (even to the extent that it is rigidly determined by the genotype) are different variables and are not perfectly correlated.

Most of the considerable confusion still remaining results from the haziness or absence of definition of the items, differences between which are to be called variations. From every point of view there is an essential difference between variation of individuals within a group and variation between groups, but the two are often inadequately distinguished. Natural selection, for instance, acts on both, but its action on intergroup variation can produce nothing new; it is purely an eliminating, not an originating, force. Despite its critics, the action of natural selection on intragroup (or interindividual) variation is essentially an originating force: it produces definitely new sorts of groups (populations), and the interbreeding group is the essential unit in evolution. Action on intra-individual variation also occurs, but, again, can only eliminate, not originate, types of individuals or of individual reactions.

There has also been some, but less, confusion of horizontal and vertical intergroup variation, that is, between contemporaneous and successive differences between groups. Without very explicit redefinition, the extension of the term "variation" to modifications of phyla in the course of time can only be confusing.

The so-called "law" of progressive reduction of variability has given rise to some thoroughgoing examples of confusion in terms and concepts. In its classic formalization by Rosa (1899) "variability" in this connection was vaguely defined, but it was applied chiefly to the capacity of a given phylum to produce new structural types. Rosa's "effective variability" was, for the most part, successive intergroup variation. Thus, by "reduction of variability" Rosa meant primarily restriction in the number of possible directions of structural modification, not what more recent students would be likely to understand as literal reduction of variability. The phenomenon is less confusingly labeled "reduction of adaptability with increase in specialization," a leading paleontological and evolutionary empirical principle today, as it already was long before Rosa.

In following passages, unless otherwise specified, the terms "variability" and "variation" will be applied to the potentiality and the reality, respectively, of differences between individuals within a possibly or actively interbreeding population at any one time.

It is well known that evolutionary change can occur without the introduction of any new hereditary factors. Aside from the influence of extrinsic factors on phenotypic variability, which is immediately reversible and hence of less long-range significance, the store of genetic variability in a large, widespread population can be unequally distributed among descendant units, which thus come to differ from their ancestors and from other contemporaneous descendants. In secular shifts within one population three related phenomena may occur: hidden genotypic variations may become phenotypic (e.g., by spread of homozygosity of a recessive), the proportions of the different variants in the population may change, and some variation may be eliminated. The latter is true reduction of variability, as the words are used here. It will be shown that apparently this is not a normal long-range phenomenon of phyletic evolution and that there is, under these definitions, no "law of progressive reduction of variability." In the differentiation of one population into several, the same processes occur, but the typical result is the unequal distribution of ancestral variants among the descendant groups. Thus, part of the intragroup variation becomes intergroup variation, and its status in and effect on further evolution becomes radically different.

The reality and importance of these phenomena have been abundantly illustrated both by experiment and by observation in recent animals and plants and hardly need further emphasis in this field. Paleontological evidence is rare and less clear, largely because of the limitations of the record and because of the deficient techniques of "standard" paleontological practice, but it can be found. Thus, I have elsewhere (Simpson 1937c) discussed an example of fossils, collected at exactly the same horizon and locality and almost certainly belonging to one population of the extinct notoungulate mammal *Henricosbornia lophiodonta*, which showed extraordinary structural variability. Some of the differences between individuals within a single group of these primitive animals are analogous to and some homologous with differences segregated in other and in allied more advanced types and then characterizing[2] distinct species, genera, or even families.

An example of a different sort—less clear-cut and open to alternative

[2] But not alone and in themselves defining.

explanation, but highly suggestive—can be extracted from Brinkmann's data (1929) on the evolution of the Jurassic ammonite *Kosmoceras*. Within the stratigraphic span of his rich and essentially continuous sequence of materials, one species, *Kosmoceras castor*, is slowly transformed into another, *K. aculeatum*, and also gives rise to a second species and lateral phylum, *K. pollux*, which in turn is transformed without further branching into another species, *K. ornatum*. The *castor-aculeatum* line is more abundant and longer lived and may be considered the main stem from which the *pollux-ornatum* line branched. I have analyzed Brinkmann's data on the following important characters: (1) terminal diameter; (2) greatest diameter of outer whorl; (3) diameter of umbilicus; (4) number of inner ribs on outer whorl; (5) number of outer ribs on outer whorl.

In all these characters except the last, which distinguishes *pollux* by its sharp progressive reduction, the earliest individuals of *pollux* are within the probable range of variation of immediately antecedent and ancestral populations of *castor*. But the variation, as measured by the Pearsonian coefficient of variation ($V = 100 \times$ standard deviation/mean) was less for all five characters of *pollux* soon after its origin than in the *castor-aculeatum* line either then, earlier, or later.[3] The amount of variation within the more abundant *castor-aculeatum* line shows only random fluctuations. Only a little later, about 140 cm. in terms of thickness of strata, the coefficients have all increased in the *pollux-ornatum* line and have become comparable with those of the continuing ancestral stock. The pertinent data are shown in Fig. 7.

In terms of evolutionary processes the most probable (although not the only possible) interpretation of these facts appears to me to be as follows: From a small section of the great, far-flung *castor* population a less abundant group was cut off and became morphologically differentiated, in part by minor qualitative genetic change, not pertinent at this point, but mainly because it received only part of the store of variability in the main population. It therefore showed at first considerably less variation than the latter. Subsequently it re-established the amount of variation usual in these groups, but did so about different means. The remaining, greater portion of the population was

---

[3] Individually the differences between the V's for the two lines at the same time are not statistically significant, but the consistent relationship for all five characters may safely be called significant.

sufficiently large so that (statistically) random withdrawal of variants into the branch unit did not wholly eliminate representation of these in the main fraction or permanently change its equilibrium, so that its variation was not noticeably affected.

Evolution by segregation of variants, otherwise expressible as expenditure of variability or as transformation of intragroup to inter-

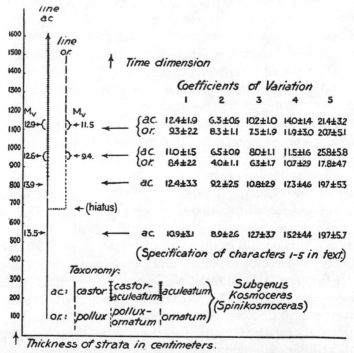

FIG. 7.—Variation in a branching phylogeny. Data from Brinkmann, rearranged. M_v, mean of the five coefficients of variation for each phylum and level. There is no abscissal scale. For further explanation see text.

group variation, is a strictly self-limiting process. In genetic terms its greatest possible theoretical limit (obviously never reached) would be the case of a completely heterozygous population giving rise to two groups, each homozygous for completely different sets of alleles. Such an extreme shift would certainly produce a great evolutionary change, but not nearly so great as innumerable changes that actually have occurred in the history of life. Obviously this cannot be the only factor in evolution, and the evidence is that it is not even important above low taxonomic levels, approximately the level of speciation.

In passing it should be mentioned that this process, like most other processes that are known to occur (and several that are not certainly known to occur) has been hailed as the dominant or only essential factor of evolution. For instance, Clark (1930) maintains that an essentially similar process is the only mode of evolution within the major phyla, and he goes so far as to state that "all mutants arise through the subtraction of something from the usual form", a curious conclusion in view of the fact (as it seems to me) that this is one mode of evolutionary change that does not necessarily involve some sort of mutation. Lotsy's theory (1916) of evolution by hybridization is another allied generalization, with the *petitio principii* explaining the origin of hereditary differences by saying that their determinants have always existed and that the succession of life as we see it results merely from their shuffling and dealing.

Evolution by segregation of previously existing variability would be like a degenerating kinetic system being limited by and tending toward a condition either of cyclic repetition or of dead level, all variability having been segregated into isolated invariable groups (transferred to the intergroup level from which no new group could be differentiated). The paleontological record as a whole shows that no such cycles have as yet appeared in the history of life on the earth. At least one paleontologist, Broom (e.g., 1933), has maintained, on other and mainly metaphysical grounds, that evolution has now almost reached a dead level and cannot progress farther; but (as he notes) few students agree. From another point of view this is a generalization, probably fallacious, of the previously mentioned principle of loss of adaptability through specialization. It does not seem necessary now to belabor these or several other theories concerned with variability that have been adequately refuted or have never been taken very seriously except by their authors.

On levels where it is a determining factor high genetic variability (as distinct from high mutation rates) is doubly significant. In the first place, it makes possible evolution by mere random segregation and loss of characters. How rapidly this can occur depends in large part upon the size of the population, and theoretically it can be predicted and evaluated (e.g., Fisher 1930; Haldane 1932; Wright 1931a). In the second place, high variability in a population may be considered a sort of bank in which mutations are on deposit, available when needed without waiting for their occurrence *de novo*. Neutral or disadvantageous characters may thus be stored up (especially, but not only, if they are recessive) and may become advantageous if a change in the nature of selection occurs. Their immediate availability then makes possible a more rapid shift in the population, a higher rate of evolution, than would be likely by mutation alone. This is one aspect of the phenomenon of preadaptation.

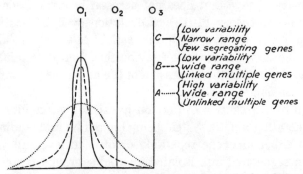

Fig. 8.—Variability, range, genetic structure, and adaptability. Modified from Mather. For explanation and discussion see text.

Recently Mather (1941) has shown how such immediately available variability can be retained without sacrificing good adaptation, in other words (contrary to Rosa's "law"), how a population may be specialized and adaptable at the same time. Polygenic combinations, such as apparently govern most essential organic structures, may be so balanced as to maintain the population mean at its optimum but continuously to preserve variation in both (or all) possible directions. As summarized in Fig. 8, curve A shows high variability and a wide range, produced (according to Mather) by unlinked multiple genes. Curve C shows low variability and narrow range, produced by reduction in the number of segregating genes (i.e., by specialization by actual loss of inadaptive variants, true reduction of intragroup variation). Curve B shows low variability, but wide range, produced by multiple linked genes.

Under strong selection pressure and relatively constant optimum (centripetal selection, discussed in a later section), a population conformant with C will be best adapted and will have the advantage over the other two types. In these conditions B also has a decided advantage over A, even though genetic recombination constantly throws off a relatively small number of more extreme and hence maladjusted individuals, who, as Dobzhansky (1938) says, must suffer for the good of the race. On the other hand, if the optimum oscillates, as it frequently does, or moves rapidly from one position to another, A will have at least a temporary advantage over B or C. For instance, as long as the optimum remains at O_1, C is the favored population. If it shifts to O_2, however, C must either become extinct or develop genetic characters wholly new for it, which is likely to be a relatively slow process. However, A will immediately have more individuals than B or C already at the new optimum and also will be able to shift its mean most rapidly to this point, a double advantage; B need not become extinct, but it is at an immediate disadvantage compared to A and also will be slower to shift the mean to the new position. Once achieved, however, its adaptation will be better than that of A, and if change from O_1 to O_2 does not occur faster than the mean of C can shift, this will be the most successful type.

If the optimum should finally shift to O_3, all three types of population must either become extinct or must develop wholly new characters; and here, again, the limiting of these processes to possibly rapid but inevitably rather small changes is seen. For sustained evolution and for larger structural changes the amount of variability existing at any one time can have little significance, except that it may be an indication of mutation rates or other more directly important factors. Concentration on low taxonomic levels, such as are alone available in experimentation, has perhaps led some geneticists to overemphasize the role of variability and to assume a too constant and direct correlation of this with the rate of evolution.

Even though their characters evolve far beyond earlier limits of range, evolutionary sequences in the fossil record do not, as a rule, show depletion of variability or progress by expenditure of variation. Moreover, although the data are not as abundant as is desirable, I can find no good evidence that the more-variable groups have, on the average, evolved more rapidly than the less-variable. These statements may be simply exemplified and may be shown to be consonant with genetic theory, although contrary to conclusions of some geneticists.

Brinkmann's data on *Kosmoceras*, already discussed, also bear on this question. In the main *castor-aculeatum* line the coefficients of variation themselves vary, but they show no significant trend and no evidence of secular depletion of variability. In the *pollux-ornatum* line the evidence suggests unusually low variability at the time of origin and a secular *increase* in variability, until it is built up to about the level normal in the other branch.

In the horse sequence illustrated from a different point of view above, the coefficients of variation shown in Table 11 were found.

TABLE 11

COEFFICIENTS OF VARIATION FOR PARACONE HEIGHT AND ECTOLOPH LENGTH IN FOUR SPECIES OF HORSES

	V of Paracone Height, Absolute Value of Which Evolves from $M = 4.7$ to $M = 52.4$	V of Ectoloph Length, Absolute Value of Which Evolves from $M = 8.2$ to $M = 20.8$
Hyracotherium borealis	6.2 ± 1.3	5.7 ± 1.2
Mesohippus bairdi	4.8 ± 1.1	4.6 ± 0.9
Merychippus paniensis	5.9 ± 1.2	5.3 ± 1.0
Neohipparion occidentale	4.6 ± 1.5	5.3 ± 1.7

Despite the great amount of evolutionary change in both these characters, their variability remains about the same throughout, with only random fluctuations as far as the data show. Moreover, the demonstrated acceleration of evolutionary rate in about the *Merychippus* stage is not accompanied by any significant increase in variability.

These figures are typical for linear dimensions of functional hard parts in mammals, for which the great majority of such coefficients lie between about 3 and 10, as further exemplified by the following examples taken at random from the many available.

Analogous variates in widely disparate mammals:

Didymictis protenus (creodont), length M_2	5.2 ± 0.7
Notostylops murinus (notoungulate), length M^3	7.1 ± 1.6
Ptilodus montanus (multituberculate), length M_1	5.7 ± 1.4

Homologous variates in ordinally related (condylarths) but otherwise very different mammals:

Phenacodus primaevus, length M_2	4.5 ± 1.0
Haplomylus speirianus, length M_2	6.7 ± 1.0

Variability does, of course, fluctuate, but for analogous characters of animals even distantly related it appears as a rule to be remarkably circumscribed in its fluctuations in time or differences in contemporaneous groups. Exceptionally low values usually indicate a sample more uniform genetically than is the whole interbreeding population, and exceptionally high values usually indicate either characters for which the mechanics and function have been less rigidly integrated or characters that are degenerating and have lost all function. This last point, particularly important, has probably become apparent to every paleontologist who has handled large collections. For instance, a functionless tooth, such as P^2 in *Hoplophoneus*, may vary from a well-developed state to complete absence, not only within one race but also within one individual (i.e., the left side may differ from the right side). In *Ptilodus montanus*, in which functional teeth have much the same variability as in other mammals (as exemplified above), the length of P^3, which is degenerating and losing function, has the high value $V = 18.5 \pm 2.8$. It is so commonly true that degenerating structures are highly variable that this may be advanced as an empirical evolutionary generalization.

Another important line of evidence would be the variability in living groups known to have had exceptionally fast or slow rates of evolution; but as far as I know little attention has as yet been given to this point. As a preliminary check, I have examined samples of several low-rate vertebrate groups, including crocodiles (little changed since early or middle Cretaceous), opossums (since late Cretaceous), armadillos (since late Paleocene), and tapirs (since about Oligocene). Full analysis is not completed, but it has been established that these slowly evolving forms all show variability at least as great as in allied, more rapidly evolving lines (e.g., lizards, sloths, kangaroos, and horses, respectively) or in reptiles and mammals in general. There is even some evidence, which requires further checking before it can be definitely accepted, that conservative groups are sometimes exceptionally variable. For instance, in a very homogeneous sample of opossums, V for tail length is 15.4, while in a similar sample of the more rapidly evolved group of white-footed mice it is 5.0, a figure usual for mammals with undegenerated tails.

Man must be considered zoologically a mammal that has evolved at more than average rate, and, moreover, he has breeding habits and other characters that must tend to increase variability. It is therefore pertinent that the variation in his homologous structural characters is not noticeably greater than in allied groups or in mammals in general. Pearl (1940, pp. 356-359), in his compilation of coefficients of variation in man, gives 70 values of linear dimensions of nonpathological groups. Sixty-three of these V's lie between 3 and 10, and 45 of them between 3 and 6. The greatest value given, 18.99 (for neck length in the Swiss) and the least, 2.35 (internal maximum length of skull in male Australians), are still within the range of V for analogous characters of more slowly evolving mammals, and the whole distribution demonstrates quite usual variability for a mammal, similar to that in single races of horses, cats, and many other groups of about average evolutionary rates within the Mammalia.

On theoretical grounds, accepting the usual genetic theories, extremely high or maximum rates of evolution would probably be accompanied, not by high variability, but by exceptionally low variability at any one time. This may be demonstrated by a hypothetical case, which could not, indeed, occur in nature, but represents the limiting extreme for rate of evolution depending on selection of favorable mutations. Suppose that a series of progressive mutations occurs in a gene: A_1, A_2, A_3, A_4, and so forth. Suppose that one such mutation appears in each generation and that the selection value for each is 100 percent, i.e., that only those animals inheriting the most progressive mutations of one generation survive to breeding age in the next. Then each successive mutation in the chain will survive for three and only three generations. The F_3 generation will be entirely different from the parental generation with regard to this gene, which (for bisexual reproduction) represents the most rapid possible evolution of the gene. Moreover, in each generation the population will be almost absolutely invariable in genotype: there will be only two genotypes, one possessed by the population as a whole, the other by a single individual, as in Table 12.

Table 12
Genotypes in Successive Generations

Generations	Genotypes	
	Mutant Individual	Rest of the Population
F_1	$A_1\ A_2$	$A_1\ A_1$
F_2	$A_1\ A_3$	$A_1\ A_2$
F_3	$A_2\ A_4$	$A_2\ A_3$
F_4	$A_3\ A_5$	$A_3\ A_4$
F_5	$A_4\ A_6$	$A_4\ A_5$

On the gene theory, continued progression of a population over long periods of time must, in more complex and necessarily slower form, approach the model of this limiting case. Earlier mutations must be eliminated from the population on one side as new progressive mutations are added on the other. Mutations are, so to speak, being run through the population, and the number present at any one time is a determinant of the variability of the population. In a large population the number of alleles actually present at any one time may be relatively independent of the rate of their loss and addition as long as these tend toward a balance and the variability pattern itself is subject to selection or some other control. This is the situation suggested by most paleontological series from large populations with moderate evolutionary rates. Decreased loss, e.g., by relaxation of selection on nonfunctional characters, would tend to increase variability, and this, too, agrees with observation. In small populations, in which the most rapid evolution is possible, rate of evolution would tend to be inversely proportional to variability, because more rapid "running through" of mutations would lower the number of alleles present in the population at any one time and slower transformation would increase the number.

The conclusions drawn from the preceding discussion and examples may be summarized in the form of the following theses on variability and evolution.

Segregation or selection of intragroup variability can give rise to new groups at a potentially rapid evolutionary rate. This process depletes variability either by its definite elimination or by its transfer to the intergroup level, where it cannot provide materials for the origin of new groups. This process is important and typical in speciation and in lower levels of differentiation.

Determinants of Evolution

- The extent and nature of variability are themselves important group characters subject to natural selection and other evolutionary factors.
- Under the influence of strong selection changing in direction, balanced polygenic combinations can produce populations well adapted or specialized, without sacrificing variability available for rapid speciation.
- Evolution on the basis of existing variability is a self-limiting process that cannot proceed beyond about the specific level.
- In continuous phyletic evolution of large populations at moderate rates variability tends toward a relatively constant level, and loss of hereditary variation is balanced by the appearance of new mutations.
- Extreme variability is more likely to occur in degenerating than in rapidly progressing structures.
- Progressive loss of variability does not appear to be a usual process or general principle of evolution.
- Rate of sustained evolution normally shows little or no correlation with variability. In more extreme cases of very slow or very rapid evolution this rate may tend to be negatively correlated with variability, but this has not been surely established.

MUTATION RATE

Discussion of the effects of mutation rates on tempo and mode of evolution must be still more theoretical and less satisfactory than discussion of other possible factors, because the factual data are deficient on both sides and the relationship between genetical and paleontological data is even more obscure than usual. Normal mutation rates in wild populations are almost unknown, although a few estimates have recently been made (e.g., in Wright, Dobzhansky, and Hovanitz 1942). The considerable body of experimental data refers to very few organisms, to relatively few of the mutating units of the organisms, and probably only to a small proportion of the mutations that actually occur in those units.

The recorded rates are, in general, those of mutations producing a well-marked discontinuous effect on some single phenotypic structure, and there is reason to believe that mutations of this sort are the least important for long-range evolution. Such mutations as most of those studied in *Drosophila* are seldom observable, or, at least, seldom definable in fossils, and when they are it is often apparent that they had nothing to do with subsequent evolution of the group concerned. Thus Osborn (1915, p. 217) has noted the symmetrical appearance of a

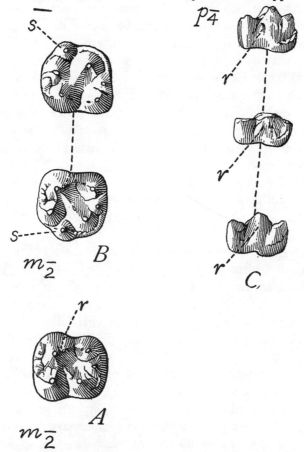

Fig. 9.—Continuous and discontinuous phenotypic variation in a fossil mammal: *A*, normal second left lower molar of *Phenacodus*, an early Eocene condylarth, crown view; *B*, right and left lower molars of an individual of *Phenacodus* (Amer. Mus. No. 15274), with symmetrically developed supernumerary cusps, *s*, presumably due to a single mutation with relatively large, definite, discontinuous phenotypic expression—"mutation" of Osborn; *C*, external views of right fourth lower premolars of different individuals of *Phenacodus* showing a region, *r*, in which there may be no cusp (lower figure), or a well-developed cuspule (upper figure), or any intermediate stage (e.g., middle figure)—"rectigradation" of Osborn. After Osborn 1915.

new tooth cusp in an individual of *Phenacodus*, which almost surely represents a new unit mutation. Osborn's original figure, despite its great interest for theoretical paleontology, was not labeled or discussed.

It is here reproduced with the pertinent data (see Fig. 9). This mutation led nowhere: it does not appear as a normal character in any known descendants of this group or even in its distant relatives. On the other hand, in some cases it can be shown that cusps that did become integral parts of a normal tooth pattern (i.e., that characterize a whole population) have arisen by minute, practically imperceptible, steps, and as far as I know there are no cases in which it can certainly be shown that such cusps have arisen by a single mutation (although this remains theoretically possible). Observations like these led Osborn (e.g., 1927) to the conclusion that "mutation is an abnormal and irregular mode of origin, which ... is ... a disturbance of the regular course of speciation."[4]

With regard to the question of size and nature of mutations, such observations will be discussed in another section. They are mentioned here to point out that it is quite impossible to count the mutations that occur in the history of any group of animals and that there is therefore no direct observational approach to the study of mutation rates during that history.

The indirect approach to this study is through variation, since by definition heritable variations arise by mutation (whatever mutations may be). From what has been said about variability, however, it is clear that variability does not itself vary directly with mutation rate. It is a complex result of many factors, and one of them, the mutation rate, seems to defy objective analysis at present. Variability could be high with low mutation rates and low with high mutation rates, and study is more hopefully directed to variability for its own sake, directly observable and a primary factor in evolutionary rates and modes. It can only be assumed, as theoretical background, that great variability implies an indeterminate but significant minimum rate of mutation.

---

[4] This statement has been unfairly ridiculed by geneticists who apparently did not know the history and some of the possible connotations of the word "mutation." By "mutation" in this sentence Osborn means only one kind of genetic mutation: a single mutation with large, unitary, discontinuous phenotypic effect. Polygenic fluctuations and all sorts of small mutations were given different theoretical names by Osborn and, indeed, are not properly mutations, in paleontological terminology, except since this has recently been modified by the less precise and historically less valid usage of the geneticists. With these semasiological difficulties cleared up, Osborn's statement seems to me true in most cases and in accordance with the opinions of modern geneticists (aside from Goldschmidt and a few others).

Determinants of Evolution

It is a common paleontological observation that some groups of animals go through stages of relatively sudden diversity, sometimes intragroup, as in the previous example of *Henricosbornia*, but more often intergroup. Among many examples of the latter sort, that of Bulman (1933) is characteristic: he shows in graptolites periods of great increase in numbers of species (intergroup diversity), followed by a sharp decline. The periods of increase show some correspondence with their attainment of new structural grades. Such periods of expansion, a sort of flowering season, have occurred over and over in the history of life among various groups successively. It is frequently assumed that these are, if not caused by, at least accompanied by high mutation rates. The assumption is reasonable, but it is no more than an assumption.

Any attempt to judge whether observed mutation rates are consistent with observed evolutionary rates involves so many untestable hypotheses and subjective estimates that little sober scientific value can be claimed for it. Nevertheless, such an attempt may be warranted as an interesting digression, and it might reveal any really gross discrepancies. Postulates relatively unfavorable to pronounced genetic change in many steps may be compared with observed morphological evolution, for instance in the horses. From *Hyracotherium* to *Equus* there must have been at least 15,000,000 generations. In the region of the direct, phyletic line, which is evidently North America, the average number of individuals actually or potentially interbreeding during this time can hardly have been less than 100,000, giving a minimum of about 1,500,000,000,000 individuals in the real and potential ancestry of the modern horse ("potential" ancestors being mainly those whose hereditary contributions were bred out by natural selection). Let it be postulated that each tooth character, for instance, is controlled by a single gene; this is the minimum postulate, since polygenic control will give more phenotypic steps per single gene mutation. Laboratory evidence (e.g., summary in Dobzhansky 1937, pp. 34-38) suggests that .000001 is a moderate or low mutation rate for relatively large mutations, while small mutations are probably much more frequent. It is thus probable that there were at least 1,500,000 new mutations in the ancestry of *Equus* since the beginning of the Eocene, as a minimum for a single genetic locus active in the evolutionary process.

Most of the mutations in the postulated situation could be only in one of two directions: an element of tooth structure can appear or disappear, become larger or become smaller. For such characters half the mutations, if they were completely random, would be in the direction favored by selection or actually followed in the morphological progression of the phylum. Setting the estimate at one-fifth seems a minimum. There would, then, be at least 300,000 favorable mutations per genetic locus in this sequence. If one-thousandth of these represented new structural changes, the actual number of determinants for progressive steps in the sequence would be 300.

If the change in any one character from *Hyracotherium* to *Equus* is divided into 300 steps, these steps are almost imperceptibly small and are almost incomparably less than the amount of intragroup variation at any one time. For instance, ectoloph length on upper-cheek teeth showed a maximum change from about 8 to about 40 mm. Divided into 300 steps, the average change in each step would be .1 mm.—the thickness of some grades of notepaper. Even in the smallest species, the intragroup range of this character was about 3 mm. Although not all measurable in the same units, other changes, such as in shapes of cusps, rise of new crests, and so forth, would be relatively even smaller when divided into 300 steps.

Under these postulates it is clear that mutations with extremely small phenotypic effects that occur at moderate or low rates, such as have repeatedly been found in experimentation, are consistent with rates of evolution not only as fast as but also considerably faster than those occurring in the record of the evolution of the horse. In general, as far as the very inadequate present knowledge goes, it seems unnecessary to postulate any marked increases in mutation rates to account for such rates of evolution as can reasonably be inferred from the paleontological record.

On the other hand, it has often been suggested in various terms (e.g., by Ruedemann 1922) that some exceedingly low observed rates of evolution imply analogously low rates of mutation or are inconsistent with what the scanty data suggest are normal mutation rates.[5] The problem of low-rate evolutionary lines is discussed later. In this

---

[5] In this connection there is an interesting suggestion (Fenton 1935) that *Drosophila*, from which the greater part of the data on mutation rates is derived, may itself have one of these abnormally low evolutionary rates.

Determinants of Evolution

section it suffices to point out that rates of random mutation in a moderate or large population do not set any effective minimum rate of evolution, although they do theoretically set a maximum. If all other conditions are conducive to stability at some fixed optimum, higher mutation rates than any yet observed would not necessarily produce any evolutionary change. Thus, when low evolutionary rates are observed, the occurrence of these conditions may be postulated, and there is no necessary discrepancy with a postulate of normal mutation rates.

These considerations, inconclusive as they are, warrant the general inference that most evolutionary sequences are consistent with moderate mutation rates and that their tempo and mode are not necessarily or primarily controlled by mutation rates. There are, nevertheless, two theoretically possible special situations to which this general inference would not apply. In the first place, it is at least conceivable that the mutation rate might become so large and other factors so weak as to make mutation the controlling factor in evolution, despite the evidence that this is not usually true. Wright (e.g., 1940, pp. 173-174) has shown that such a situation may arise in very small populations. With mutations random relative to the optimum adaptive adjustment, only a small proportion are likely to be adaptive, less than half in any typical case. Hence, as Wright points out, the systematic effect of mutation under these conditions would be degenerative. Still more important, and to be considered below, are the facts that the direction of evolution would as a rule be inadaptive and that in a small fraction of such populations it would be pre-adaptive for some adaptive type and level quite unlike those ancestral for the group. In these few, but highly important, cases the rate of evolution would tend to vary with the rate of mutation and the direction or nature of evolution would also be influenced by that rate, being more adaptive with lower and more degenerative and pre-adaptive with higher mutation rates.

The other special circumstance in which the mutation rate might become a primary factor of evolution would arise if mutations were strongly directional and not random, even in a limited sense. Thus, as the simplest example, if all mutations tended toward larger gross size in, let us say, the proboscideans, and none smaller, then increase in size would be almost inevitable and the rate would vary more or less clearly with the mutation rate, unless this increase were so inadaptive that it would be entirely prevented by selection (when the group would cease to evolve or would become extinct). The phenomenon would be one of momentum and will later be discussed as such. The reality and wide occurrence of such situations are often claimed on the evidence of so-called orthogenetic paleontological sequences. Evidently the situation could arise and perhaps it has, but I know no instance that is explicable only in this way, and I believe that this has been most exceptional in evolutionary history. Without taking up all the reasons for this opinion at this point, it may be noted (as Osborn, for instance, strongly emphasized, but interpreted differently) that "orthogenetic" evolution is almost invariably adaptive. When change in conditions makes another direction adaptive, the direction does change accordingly, e.g., proboscideans on islands, where smaller size became advantageous, reversed the previous trend and did become smaller.

CHARACTER OF MUTATIONS

Probably the most disputed evolutionary question, both in genetics and in paleontology, today has to do with the nature of mutations, both in general and in the specification of the particular sort of mutation involved in a given evolutionary process.

The word "mutation" was first applied by a paleontologist, Waagen (1868), to a recognizable stage in a continuously evolving phylum, hence to a population and to a taxonomic unit, more or less of subspecific rank in terms of morphological definition, but differing from a zoological subspecies by being vertically (temporally) delimited rather than horizontally (geographically or ecologically). Then De Vries (1901) found that new types of organisms, comparable to new taxonomic groups, sometimes arise all at once and to this very different phenomenon he applied the same word "mutation." Then Mendelian genetics came to use De Vries's term in a more exact sense, meaning a change in phenotypic expression associated with some one locus in a chromosome and segregated according to determinable rules. Then it was learned that similar phenotypic changes may be related, not to such a locus, but to various aberrations in a whole chromosome or in the set of chromosomes, and these, too, are often called "mutations," or "chromosome mutations," as opposed to "gene mutations." Difficulties in maintaining these distinctions have finally led to the

Determinants of Evolution

frequent use of the word "mutation" to mean the appearance of a character not inherited from the ancestors of the organism in question, but heritable by its descendants, regardless of the exact cytological localization or physiological mechanism of the change.

Some paleontologists (e.g., Swinnerton) long continued the use of "mutation" in the sense of Waagen and others (e.g., Osborn) in the sense of De Vries. The geneticists, many of whom seem to be unaware of earlier meanings of the word, have long since abandoned these usages, and it would be merely confusing to continue them now. Unless otherwise specified, it is safe only to assume that the word is used in the latter, most general genetic sense.

The old argument whether inheritance is particulate or blending is not wholly settled, but it has been conclusively demonstrated that it is particulate in some cases, and most students believe that this is a valid generalization. Room for doubt has remained because there seems to be an essential distinction between manifestly particulate expressions of inheritance, segregating Mendelian factors, and others that do not clearly segregate or do, in fact, blend. Further genetic investigation proposes to solve this dilemma by demonstrating that inheritance, the actual set of germinal factors passed from one generation to another, may be particulate even when its expression or phenotypic results are not, the blending being explicable by such things as polygenic characters, modifiers, conditional dominance, ontogenetic factors, and so forth. Goldschmidt (e.g., 1938) and others have further clarified the situation by emphasizing that no organic structure or function is inherited as such, but only a system of determiners for their chemicophysical development.

Paleontologists, particularly, have been much disturbed by the problem whether the rise of new characters is continuous or discontinuous. Some maintain that the process is discontinuous, thus Spath (1933):

We know now that new types appear as saltations and that it would be as idle to insist on seeing a complete series of passage forms from *Orthoceras* to the belemnites, as a progressive modification of the shell in a baculite or other hamitid.

At least as many paleontologists, probably more, hold the opposite opinion, despite Spath's "we know," thus Matthew (1926):

The more complete the record, the more abundant the material, and the nearer we are, judging from the available evidence, to the probable centre of evolution and dispersal of any race of animals, the more continuous does the succession become, the more it appears to evolve through a succession of minute changes which lie within the limits of ordinary individual variation.[6]

In its crudest form, the distinction between continuity and discontinuity in evolution is almost meaningless. The developed organism is absolutely discontinuous from its parent (with exceptions not modifying the conclusion), and any real difference between the two, however small, is discontinuous; there is no morphological continuum. Part of the essential point is, rather, whether the discontinuity, which arises by mutation if it has a heritable basis, is large or small. The question has also been much discussed in this form, but still with some confusion, because "large" and "small" are grades on a continuous scale and can be defined only subjectively. There are many characters, e.g., number of vertebrae, that cannot vary continuously and for which there is therefore a minimum "size" of mutation that would be considered "large" if occurring in a continuous variate, e.g., size of a tooth cusp. There is no good basis for belief that a large mutation, as expressed in one character, differs in kind, as well as in degree, from a small mutation, but in general terms it is possible to make the quantitative distinction and to investigate its effect on evolution.

Matthew implied a more important distinction in the passage previously quoted. Expressed in different words, this depends upon whether the differences between successive populations (successive intergroup variation) transcend the limits of differences within a single population (intragroup variation), however broad or narrow those limits may be. This, I think, is the most valid form and the most fruitful basis for discussions of the problem more confusingly labeled continuity and discontinuity.

Another disputed distinction, made in Goldschmidt's recent work (1940), is between mutations that have one or a few specific phenotypic effects (e.g., most of the studied gene mutations) and those that

---

[6] Of course Matthew did not deny the obvious fact that sharply discontinuous changes can occur, at least in single characters. He meant only that they are not the usual stuff of evolution and that they are relatively and absolutely unimportant in known phyletic series. The same conclusion was that intended by Osborn when he called mutations mere accidents interrupting speciation, as quoted on a previous page. Matthew distinguished mutations (i.e., in the sense of De Vries) from "minute heritable variations," and his work is confusing to geneticist readers, despite the great clarity of his mind, because he also habitually used "mutation" in the sense of Waagen.

affect the whole organism and set up a new genetic system, the systemic mutations of Goldschmidt. The distinction seems to me largely an artifact resulting from laboratory procedure. Most of the mutations studied are those that produce sharp phenotypic modifications and are usually labeled by an effect of this sort on some single structure; but in fact most single genes clearly influence more than one structural element of the organism, and the effect of a single gene mutation may be both local and general or may be altogether general, i.e., systemic (see Dobzhansky, Waddington, or almost any recent textbook on genetics). Moreover, the chromosomal aberrations, claimed by Goldschmidt as the basis for systemic as opposed to gene mutations, have phenotypic results indistinguishable from those of gene mutations, or supposed gene mutations (see Stadler 1932), and, like the latter, both local and general phenotypic effects.[7] In other words, anatomical units and genetical units do not coincide, as is now well established and generally recognized, despite the fact that the practical exigencies of laboratory procedure and nomenclature can lead the unwary to assume coincidence.

Multiple and systemic effects of many gene mutations are now commonplaces to the geneticist, but the combination of such effects is often surprising, and their frequent lack of relationship to the "unit characters" of morphologists and taxonomists needs repeated emphasis. An example of unusual interest to paleontologists and mammalogists is given by Snell (1931). "Short ear" in the house mouse, established as a single mutant gene, produces not only the character for which it is named but also a significant reduction in palatilar length and in rostral height of the skull. Thus, one mutation is found to affect three different "unit characters," each of which is commonly accounted an important and independent taxonomic distinction, one an external character universally used to define groups of living mammals and two skeletal characters widely used in paleontology.

The most important evolutionary problem involved in studying the

[7] It is a fallacy running through Goldschmidt's work that he assumes and his argument demands a clear-cut dichotomy between micromutations, particulate, small in phenotypic effect, at loci in chromosomes, and macromutations, systemic, large in phenotypic effect, caused by chromosomal aberrations. This is not one dichotomy, but at least three, independent of each other. The contrasting of characters proper to a branch of one dichotomy with those proper to a branch of what is, in reality, a wholly different dichotomy, as if these were the two alternatives in one closed system, repeatedly leads Goldschmidt to *non sequitur.*

nature of mutations is whether new types of populations, new taxonomic units of one rank or another, normally or frequently arise by saltation all at once, in one individual, by a single genetic event, or whether all such groups usually arise more gradually by the spreading within a population of single mutations less radical in effect.

Theoretically saltation might result from a single large mutation, from the simultaneous occurrence of multiple smaller mutations, from a single radical chromosomal aberration, or from the sudden segregation or recombination of pre-existing genetic characters (e.g., segregation of animals homozygous for various recessives or systematic recombination of a considerable number of dominants).

It is proved for large single gene mutations and for chromosomal aberrations that these processes do occur and that they can produce new races and species, although the known examples are few and exceptional. Races of *Partula*, as shown by Crampton (1932), can be defined by the occurrence and proportions of dextral and sinistral coiling, a very marked morphological difference controlled by a single gene. This may be called "saltation"; but it is not an example which supports the usual conception of that process, especially as it is conceived by paleontologists who believe in normal evolution by saltation. The coiling is a discontinuous variate, having only two classes: it must be either dextral or sinistral. The one-step genetic change involved is not necessarily greater than might be a change of 0.1 mm. in the mean of some dimension, which could reflect an equally great change in one gene. Moreover, although there are purely dextral and purely sinistral races, there are also intergrading races partly dextral and partly sinistral within the same breeding population: the intergroup variation does not differ in kind or degree from the intragroup variation of some groups.

Perhaps the most nearly convincing evidence for the possibility of systemic mutation, or "macroevolution" as used by Goldschmidt, is provided by the phenomenon of homoeosis, recently well reviewed by Villee (1942), whose conclusions are fully Goldschmidtian. Very briefly, the experimental facts are that single mutants may produce phenotypic characters typical of other families (or still higher groups) than that in which the mutant appears. Thus the mutant tetraltera in *Drosophila* produces wings like those of a different family of Diptera, while

Determinants of Evolution 53

the mutants proboscipedia, bithorax, and tetraptera all produce phenotypic characters of other orders of insects.

It must be noted that authorities on homoeosis do not generally agree with Goldschmidt's or Villee's opinion of the evolutionary (rather than strictly genetic) meaning of the phenomenon. Thus, Dobzhansky, who (with Bridges) described proboscipedia, finds homoeotic phenomena fully consistent with the neo-Darwinian position. There appear, indeed, to be two major fallacies, or at least imperfections, in the Goldschmidt-Villee interpretation. First, homoeosis cannot fairly be called the origin of a new "reaction system" in the sense of Goldschmidt. It concerns single mutations (whether of genes or chromosomes) producing essentially a unitary genetic effect, apparently the same in kind as the smaller mutations that are the basic creative element of neo-Darwinism. Though they are great in degree of phenotypic effect, such mutants do not change, because they may be taken as merely the extremes of a graded series. Such extremes are demonstrably much less frequent than smaller but otherwise analogous mutations, and in theory (no observations clearly opposing) they are far less effective in evolution. Tetraltera produces a *Drosophila* with one *Termitoxenia*-like character. This is very far from changing a *Drosophila* (or any other) genetic reaction system into a *Termitoxenia* system and the coincidence of a great number of such unit mutants would still be necessary to produce generic changes such as are most common in phylogeny.

The second and related objection is that the appearance of a mutant individual is not evolution. Only populations, not individuals, evolve. However profound or systemic a mutation may be, it is difficult to see how its eventual effect on populations can be independent of the neo-Darwinian factors. The sudden appearance of a really new individual reaction system morphologically of generic or higher grade would not produce a new genus or higher genetic unit unless and until this system came to characterize an isolated population. If this phenomenon could be substantiated as real, it would supplement, not supplant, the population theories that involve selection acting on unit deviations of variable but usually small degree.

Aside from the laboratory phenomenon of homoeosis and other phenomena that do not, in reality, demonstrate the occurrence of systemic mutation as a normal factor in evolution, there are exceptionally great inherent difficulties in substantiating or disproving such an occurrence in nature. The paleontological record proves that it did not occur in some particular instances, but is and always will be incapable of proving that it never occurred. Nor can this record ever give positive proof that it did occur. The most nearly positive observation of this kind in recent faunas known to me is that of the nemertean *Gorgonorhynchus* which is believed by one of its discoverers to have arisen suddenly as a new genus sometime between 1903 and 1931 in widely separated parts of the globe (Wheeler 1942). The data are quite inadequate to establish the validity of this interpretation. In the nature of things such data must perhaps always be inadequate, but it would be hollow dialectics to rule out other explanations because the facts conclusively opposing them, if any, cannot be observed. Moreover, on Wheeler's showing, almost nothing is actually known on the positive side as to the real mode of origin of this supposed genus, its affinities, or its own genetic, breeding, and population structure. For instance, among many other possibilities, it may be that the supposed new genus is simply a new or hitherto less-common morpha produced within a population by a large, but not properly systemic, mutation analogous to mutations producing homoeosis. In such a case, or in other cases equally possible for this example, to say that a new genus has arisen in one step is only a manner of speaking, and there is no real contradiction of the neo-Darwinian interpretation of the same facts. Until something is known about these other factors, the discovery is only a curious anomaly; and Wheeler's explanation of it remains a rather fascinating possibility, not a probability and most certainly not a fact.

Simultaneous appearance of several gene mutations in one individual has never been observed, so far as I know, and any theoretical assertion that this is an important factor in evolution can be dismissed. Even throughout the vast span of geologic time the probability of such an event is negligibly small. For instance, postulating a mutation rate of .00001 and supposing that the occurrence of each mutation doubled the chances of another mutation in the same cell—a greater departure from random incidence than is likely to occur—the probability that five simultaneous mutations would occur in any one individual would be about .00000000000000000000001. In an average population of 100,000,000 individuals with an average length of generation of only

Determinants of Evolution

one day, such an event could be expected only once in about 274,000,000,000 years—a period about one hundred times as long as the age of the earth. Such an occurrence obviously has not been frequent enough to take into account as a real factor in evolution, especially when it is remembered that it would have to produce a viable, integrated organism at least as well adapted to an available environment as any forms already existing and that the chances of doing so are also practically infinitesimal.

The processes of segregation and recombination are of great and obvious importance in evolution, but the chances of their producing evolution by saltation, in the usual sense, are likewise negligible. Clark (1930) apparently had some such process in mind, but he did not discuss its mechanism. According to any known or, it would seem, possible mechanism, short of divine intervention, such effects in a population fluctuate about defined points of equilibrium, and change to a new equilibrium can only be accompanied at finite and usually moderate rates largely determined by such extrinsic factors as selection pressure (e.g., Fisher 1930). Moreover, such processes cannot introduce any new genetic characters into a population, and a theory of evolution that does not account for the emergence of true novelty begs the question and is hardly worthy of serious consideration.

The virtual impossibility of the simultaneous appearance of a number of morphologically congruent random mutations and the obvious fact that different functionally related characters do evolve in unison underly the revolt by some paleontologists and others against the belief that gene mutations as they occur in the laboratory have anything to do with evolution in any broad sense. In the first place, neither the experimental nor the observational data require or warrant the belief that mutations are completely random in their phenotypic effects, although they may be nearly random in incidence (i.e., have essentially random distribution in a population and in time). On the contrary, considered as samples of an infinite number of possible morphological changes, most possibilities of mutation appear to be rather rigidly limited. For many characters only two directions of change are possible, for instance, toward larger or toward smaller size, and mutations in one of these directions may be more frequent than in the other. If the chances are about even that a mutation will be toward or against an existing selection pressure, it may still be considered random from the point of view of adaptation and functional integration. It is, indeed, unnecessary to assume that mutations are normally random even in this limited sense. In a highly specialized and well-adapted organism the chances are that any one mutation will be disadvantageous, as in *Drosophila*. In a more poorly integrated and poorly adapted organism the chances of advantageous mutation are evidently much greater. In the second place, the problem of functional integration becomes much simpler if useful mutations are very small but numerous and may, indeed, be insuperable if mutations are few and large. Finally, part of this problem of integration arises from a misconception of the particulate nature of inheritance. If, as some students have implied (e.g., Wood & Wood 1933), each cusp and each tooth were governed by a separate gene, the difficulty would indeed be tremendous or quite insuperable. But there is abundant and excellent evidence that genetic factors, even though particulate, are not invariably or commonly specific—that, for instance, the dentition as a whole is subject to field control by genetic factors not specific to any one tooth or part of a tooth (see Butler 1939). Thus, the reaction of simultaneous mutations in numerous different genes need not raise any difficulty in the concordant progression of functionally related structures.

The differences between related groups, even on the lowest taxonomic levels, usually involve a large number of different genes (Timofeeff-Ressovsky 1940). Experimental analysis of differences of high taxonomic rank is impossible, but it is safe to infer that they may involve hundreds or thousands of particulate genetic differences, each of which arose from a shorter or longer sequence of individual mutations. The preceding considerations are sufficient to induce agreement with Goldschmidt that radical chromosomal aberration is the only conceivable way in which such differences might possibly arise *by one step*. Without attempting to review, and certainly without depreciating, the large mass of factual data on the problem assembled by Goldschmidt (1940), I do not believe that these data require or are even consistent with his conclusion that these differences do arise by one step. Most of the saltations that are known to have occurred (all on the lower taxonomic levels) are commonly chromosomal aberrations, including most of the famous "mutations" of De Vries (see Emerson 1935). Polyploidy is common only in plants, and even in extreme cases the morphological change produced is not greater than between many natural species.[8] Translocation and inversion, alone or as single events, also produce not more than specific distinction in the most extreme cases, and their effect on the whole is not that expected or demanded by the theory that such is the mode of origin of higher categories. Thus, in *Drosophila* the two similar but distinct species *D. melanogaster* and *D. simulans* have almost the same chromosome arrangement (Sturtevant 1929; Kerkis 1936), while the equally similar species *D. pseudoobscura* and *D. miranda* have radically different arrangements (Dobzhansky and Tan 1936), and some nearly indistinguishable groups of *D. pseudoobscura* also differ markedly in arrangement (Sturtevant and Dobzhansky 1936). The occurrence of intergeneric hybrids (e.g., *Bos* × *Bison*) also shows by inference that there may be no important difference in chromosome arrangement between groups that differ markedly in morphology and certainly have many differences in genes.

Translocations and inversions may produce position effects analogous to and even indistinguishable from gene mutation (e.g., Dubinin & Sidorov 1935; see summary in Dobzhansky 1937). But these are, from the point of view of rates of evolution, on about the same footing as gene mutations. The distinction is important only because it may increase the probability of coincidental appearance of mutations, but it has been shown that this probability may be considerably increased without permitting instantaneous transition between higher categories. It is also important that chromosome effects probably do not and surely are unlikely to produce any qualitative distinction in the particulate units of heredity, such as are normal, if not universal, between higher categories. It is improbable that any amount of gene rearrangement or recombination can in itself produce any truly major effect in evolutionary history.

The supposed paleontological evidence for saltation as a normal mode of evolution is drawn mostly from instances of sudden breaks in the record or of the sudden appearance of new groups. These problems will later be treated in sufficient detail. The general conclusion is that the evidence does not require saltation as an explanation, but is more

---

[8] If one granted as a premise Goldschmidt's conviction that macroevolution, essentially one process, occurs on specific and all higher levels and microevolution, essentially a different process, on subspecific and lower levels, this admission might warrant his other generalizations. But this is not an acceptable premise. It is a hypothesis that must stand, or evidently fall, on this evidence that is so clearly opposed to it.

simply and more probably explained in other ways. Because of the nature of the record, it is quite impossible to prove the negative generalization—that saltation never occurs. There is, however, abundant and incontrovertible paleontological proof that saltation does not always occur, i.e., that continuity (with Matthew's meaning) or gradual intergradation commonly occurs between what are certainly good species and genera.[9]

The paleontological record is consistent with the usual genetical opinion that mutations important for evolution, of whatever eventual taxonomic grade, usually arise singly and are small, measured in terms of structural change. With regard to the experimental and theoretical data of population genetics, it has been shown by Timofeeff-Ressovsky (e.g., 1935), through experiments, and by Dubinin and others (1934), through analysis of wild strains, that small mutations are much more frequent than large mutations. Fisher (1930) has shown that the probability that any one mutation will have selective value is inversely related to the size of the mutation and diminishes from a limiting value of 0.5 as the phenotypic effect of the mutations increases. Thus, large mutations are not only less frequent than small ones but also less likely to be advantageous or even viable.

An unusual example of a large mutation, that had no useful outcome and small, useful mutations in the same stock has already been given. Osborn's work (e.g., 1934, 1936, 1942 and other works there cited) is full of incontrovertible examples of the independent development of morphological characters by small steps, overlapping between successive populations. The large body of factual evidence assembled by this great paleontologist and the validity and meaning of important

---

[9] To those who have done much work on good phyletic series of fossils it will hardly seem necessary to make such an obvious statement as that good species and genera frequently arise in this gradual way, whether or not they always do; but Goldschmidt's widely publicized work denies this fact (which vitiates his basic argument) and even claims paleontological support for this denial. In spite of assertions that he has drawn facts from all fields and that paleontology will be shown to support his thesis, the section of Goldschmidt's book that purports to give a few facts from paleontology gives no facts at all and only cites one of the thousands of pertinent paleontological studies, with the statement that this "leads to exactly the same conclusions as derived in my (Goldschmidt's) writings, to which he refers." After having carefully read the paper cited (Schindewolf 1936), it seems to me not only not to lead to some of the main conclusions of Goldschmidt but also to contradict them. This paper, although a theoretical work of great importance and value, is also in some respects, especially where it does approach some of Goldschmidt's conclusions, at wide variance with the consensus of paleontologists and even with some of its own author's other works.

deductions from this evidence have been undervalued, probably because they were obscured by Osborn's personal and peculiar terminology and by the metaphysical nature of his conclusions as to causes, which are quite independent from his purely physical conclusions as to modes.

As typical of many cases, an additional example will be briefly given in which characters having essential evolutionary importance and which subsequently become associated with others in such a way as to be definitive of generic units, arise singly and by small mutations within a population in which they are not modal, definitive, or "normal." One of the essential differences (for purposes of practical identification) between the Oligocene horse genera, *Mesohippus* and *Miohippus*, and their immediate successor, *Parahippus* is that the latter has and the former typically lack a tiny spur or crest on the upper cheek teeth, which in later members of the line to *Equus* gradually became a large, essential element in the pattern, the crochet. But even in the earlier *Mesohippus* there are occasional forms with a small crochet, found in direct association with and otherwise indistinguishable from more numerous individuals without the crochet. Evidently these are single populations in which the mutation "small crochet" had occurred, but had not yet become abundant or segregated (see Schlaikjer 1935[10]; Stirton 1940; also American Museum of Natural History collections, which confirm and reinforce the conclusion). In some earlier species of *Miohippus* the same situation obtains, and among latest Oligocene forms there is some evidence, still inadequately analyzed, that segregation was occurring and that closely related species were in part distinguished by relative constancy of presence or absence of the crochet. In the early Miocene the segregation is almost complete and constant. Some well-defined species (evidently the type of *Miohippus, M. annectans,* which is not an average or modal species in the

---

[10] Aside from specimens referred to *Mesohippus barbouri*, some of which do have, but most of which do not have a crochet, Schlaikjer set aside a few closely similar specimens from the same quarry and erected for them a different genus, *Pediohippus* (rejected by Stirton), defined principally by the constant presence of a crochet. He explained the interpretation of the supposedly distinct genera by the hypothesis that they hybridized. The explanation seems to me improbable, but I have not reexamined the specimens and Schlaikjer does not give sufficiently full data to permit critical revaluation. Even if his alternative explanation be granted, it does not alter the conclusion here reached except to claim that some of the forms with a crochet did become segregated earlier than other students believe.

genus) then apparently never had a crochet, and they gradually gave rise to later genera, e.g., *Anchitherium*, in which this structure did not develop. Other early Miocene species, e.g., *Parahippus pristinus*, then constantly had a small crochet and are, mainly on that account, placed in a different genus, having intergraded almost imperceptibly with earlier *Miohippus*, but later becoming more distinct, so that they unquestionably warrant generic separation. From these species came the later true equine horses, in which the crochet became large and important. It can be seen that other tooth characters, helping to distinguish between the typical forms of the genera, arise in the same way, but separately from the crochet. For instance, the first mutations "incipient united metaloph" begin to appear as fluctuating rarities in populations normally lacking them, probably slightly later than the first crochet mutations occur. This character, unlike the crochet, did not become segregated as a distinction between contemporaneous groups, but spread and became universal in later Equidae.

In these cases, and generally in similar paleontological data, the phenotypic expressions of mutations do not, even aside from the fact that they are smaller, have the clearcut nature of such mutations as "forked" in *Drosophila* or "sinistral" in *Partula*. Even when such

TABLE 13

OCCURRENCE OF A CINGULUM ON LOWER CHEEK TEETH OF LITOLESTES NOTISSIMUS

SEPARATE TEETH

Teeth	No. of Observations	No. with Distinct Cingulum	Percentage
P_4	24	2	8½
M_1	35	11	31½
M_2	36	8	22
M_3	24	3	12½
Totals	119	24	20

ASSOCIATED M_{1-3}

Cingulum on	Number	Percentage
M_{1-3}	1	5½
M_{1-2} only	2	11
M_{2-3} only	0	0
M_1 and M_3 only	0	0
M_1 only	3	16½
M_2 only	0	0
M_3 only	1	5½
None	11	61
Total	18	

Determinants of Evolution

characters first appear, frequently they cannot be classed as "wholly present" and "wholly absent," but already show variation both in size and in extent; for instance, in the number and location of the teeth that show them. This last point is further exemplified by the data in Table 13 (from Simpson 1937d) on the occurrence of a cingulum on lower cheek teeth of *Litolestes notissimus,* a late Paleocene condylarth. Some allied groups lack this structure, and in others it is relatively constant.

It is most improbable that such characters were arising by mutation of genes having a 1:1 relationship to the phenotypic structure. A more likely hypothesis is that the mutation involved, although the same in all individuals showing the character and quite possibly present in some that do not, introduced a new tendency in a complex developmental field and that the phenotypic expression fluctuated under the influence of other genes related to the same field and of extrinsic factors influencing development.

Among possible exceptions to this rule of fluctuating and essentially continuous origin of new characters, Robb's data on the reduction of lateral digits in the horse are among the best documented (see Chapter I). Here the whole amount of genetic change demonstrated for this one character occurred at a single stage in horse history—before this point the lateral digits were functional and about 1½ times as long as the cannon-bone, and afterward these digits were vestigial and about ¾ as long as the cannon-bone. The change seems to have occurred in one step, and the character is not demonstrated to have varied in any one population. It can be assumed, however, that the latter is merely a deficiency of record and study, and it is probable that the former is, also. There is no theoretical reason why the change may not have been accomplished by one mutation of clear-cut dominant effect, and the transition was certainly relatively rapid; but survey of the scanty intervals suggests that here, too, there actually were several small steps rather than one large step. Robb did not analyze specimens actually in the transitional period, and sufficient data on these are not at hand. If it should be proved that instantaneous change actually occurred, this will be an unusual example of a relatively large mutation involved in important progressive adaptation. It would, however, differ only in degree from the more usual process, and it would not be a saltation in the sense of the sudden rise of distinctly new taxonomic units. The lateral digit reduction was essentially independent of many other more

Determinants of Evolution

gradual and fluctuating changes, e.g., in the teeth and even in other foot characters, and the rise of new population units involved all these.

LENGTH OF GENERATIONS

The average length of generations must influence the rate of evolution. If, as is now generally admitted, the heredity of an animal is fixed permanently before the first cleavage of the developing embryo, no further evolutionary change (apart from very rare and unimportant somatic mutations) occurs until the next generation appears. It is true that some environmental factor may so modify the phenotype in a section of a population that it will become morphologically distinct from another section practically identical in genotype. It may then be said that evolution has occurred during rather than between generations, and this may be a more widespread phenomenon than is generally realized, both among plants (e.g., Turrill 1940) and animals (e.g., Hogben 1933), although some of the supposed examples probably involve a real change in the genotype produced by natural selection. It is, however, clear that the sort of evolution involved here is, from one point of view, spurious and, from any point of view, confined to very low taxonomic levels. No change in structure at high levels is reasonably explicable in this way.

In normal evolution the discrete steps occur between generations, so that the natural rate of evolution is a rate per generation, and a true temporal rate must include the length of generations as a factor. The length of generations is also a factor in the effective rate of mutation. Rate of mutation is usually expressed per individual, being, for instance, about one in a million individuals, or .000001, for a given mutation. But the actual result of mutation rate on evolutionary rate is likely to be effected not only by the relative number but also by the absolute number of times that a mutation occurs, and this depends upon the absolute number of individuals in which it could occur, which is the average size of the population multiplied by the length of time involved and divided by the average length of a generation.

All the temporal processes of evolution are more naturally measured in generations than in direct time units. The effect on all these processes should be an inverse relation between rate of evolution and length of generations. A priori, it would appear that the influence would be profound, because generations vary so greatly in length, from less than a day to more than thirty years. The fossil record, however, seems to show no close or obvious correlation between rate of evolution and length of generations. Among mammals, opossums have short generations and elephants are near the maximum; but the evolution of the elephants has been many times more rapid than that of opossums. Among horses the length of generations must have been about the same in the closely related *Neohipparion* and *Hypohippus* lines, yet in most respects the former evolved more rapidly than the latter. Invertebrates have, as a rule, shorter average generations than have vertebrates; but on the whole vertebrates have evolved more rapidly. There are so many exceptions of this sort that they cast some doubt that length of generation has any causal relationship in the cases in which short generations do seem to have accompanied rapid evolution.

In a general way length of generations is correlated with body size, and its possible influence is difficult to separate from that of other factors similarly correlated, e.g., the adaptive significance of body size itself, the extent of geographic range, or the size of populations. Although there is no clear evidence that body size has essentially influenced rates of sustained phyletic evolution throughout geological periods, it is clear that, on the average, racial or specific differentiation on a geographic basis is more rapid and more complex for small animals than for large animals. The shorter generations of small animals may be a factor in this.

Small size and correlated relative shortness of generations also deserve consideration with relation to the problem, discussed subsequently, of the apparently sudden rise of major groups, orders, classes, and so forth. Individual size in such groups almost invariably averages smaller when they first appear in the fossil record than at any subsequent time in their history. With a few possible exceptions, the individual size at the time of appearance also averages smaller than in allied or competing groups at the same time. The probable relative shortness of generations in these groups just before they appear in the fossil record would at least facilitate unusually rapid evolution at that time, although it can hardly be the principal explanation.

In a less direct way, length of generations may have another, more profound influence on evolution. Almost all organisms have to meet cyclic fluctuations or secular shifts in the environment, and the selective effect of these changes depends in good measure upon their relationship to length of generations. Thus, land animals in the temperate zone must survive both hot summers and cold winters. The many forms with one-year generations can meet the situation by structural, as well as physiological, adaptations in different stages of the life cycle, as do many insects. Others, with longer generations, must as a rule develop wide physiological tolerance, as do many mammals, or hibernate, as do other mammals, or migrate, as do many birds.[11] Some animals with short generations, notably insects, can be adapted in various stages of the life cycle to one particular food, available only seasonally. Others, with longer periods of activity at stages of the individual cycle, cannot be so specialized, but must adopt either a continuously available diet or a diet varying with the seasons. In general, the individuals of successful phyla must be capable of adjustment to environmental changes that occur during their lifetime, just as a surviving phylum as a whole must be capable of adjusting to the changes that occur during its span. The two periods involved may even be of comparable length in extreme cases. It has been observed that some populations have made genetic adjustments to changes in environment within periods of five years and even less (e.g., *Carinus,* Weldon 1899; *Aonidiella,* Quayle 1938), a period exceeded by the length of generations in many animals. The specificity, as well as the nature, of population (genetic) and of individual (structural-physiological) adaptation is profoundly influenced by this relationship.

As a first suggestion, it is just possible that this effect was involved in the differential extinction at the close of the Pleistocene, although again it was not the only factor and probably not the principal factor. Then a greater proportion of large animals became extinct than of small animals (e.g., Simpson 1931). Throughout the long Pleistocene epoch (a million years, more or less) these large forms, with long generations, would tend to make a slow genetic adjustment to the lower average temperature, or generally inclement climates, in a specific region, and readjustment to the more moderate recent climates would

[11] Similar co-ordinations of life-span with climatic cycles are even more striking in plants, with their well-marked physiological differentiation between one-year generations and longer generations (annuals and perennials). Normally the annuals meet the temperate-zone winters as do annual insects, by a dormant or protected early period in the life cycle (dormant seeds). One evidence that this is genetically controlled is the occurrence of mutant maize, in which the seeds do not have a dormant period (Eyster 1931). Deciduous perennials are analogous to hibernating mammals and, like the latter, often have nondeciduous (nonhibernating) relatives in regions of less severe winters.

be slow. Smaller animals would adjust and readjust more rapidly and might tend to adapt primarily, not to a long average environment, but to a short summer-winter swing within the life cycle, more severe, but of the same nature as the shifts in their present life cycles.[12]

SIZE OF POPULATION

The simplest genetic pattern of progressive evolution is one in which for a given gene in a population there arises by mutation an allelomorph which spreads through the population, gradually replacing the original gene, which may eventually be lost; the mutant form then becomes universal. This is not the only genetic process involved in evolution, but it is unquestionably an important factor. It can best be expressed on a scale of gene frequencies or of ratios of allelomorphs within a population. Fisher (especially 1930), Wright (especially 1931, see also later references in the Bibliography), and Haldane (especially 1932) have studied the theoretical effects on these ratios of a number of variables in different sorts of populations set up as models. The principal variables considered are mutation rates and intensity of selection. Fisher and, more particularly, Wright have also considered size of populations and their breeding structure, and Wright has further emphasized the special effects of migration between adjacent interfertile populations. Wright's latest work (e.g., 1942) comes nearest to a generalization in which the most pertinent variables are simultaneously involved, and his models nearly approach the conditions that have been observed in real populations. All the proposed models, which sometimes give different results, but agree on many essential points, serve to analyze the various factors in relatively pure form and to clarify their usually very complex interactions in nature. The correlation of these theoretical deductions with observations on real populations is under way (e.g., by Wright and Dobzhansky and their colleagues), but this is an essentially new line of research in which most remains to be done.

Some reference to these studies has already been made. They are most pertinent to the present section, for it is they that have made

[12] After this passage was first written, Eiseley (1942) made a somewhat similar suggestion with regard to the species of *Bison* in North America. He points out that the large species of the Pleistocene may have become extinct because they achieved a long-range adaptation to ice-age conditions and that *Bison bison* may have survived because it did not.

apparent the essential role of population size as a determinant both of rates and of patterns of evolution. Population size (symbol N in the studies cited) is in this connection more limited than in its vernacular sense. It refers to the genetically effective breeding population. In a group reproducing sexually, in which males and females are in about equal numbers, it is the number of individuals actually engaged in the production of offspring at any one time and so situated that any two might be ancestral to a single individual in some future generation. This is much less than a census figure for population. In a common alternative type of breeding structure in which the breeding males are much less numerous than the females, the effective value of N is about four times the number of actively breeding males (Wright), a figure lower still. In a population in which the number fluctuates the effective breeding population over a period of years is near the lowest figure reached at any one time in that period.

In large populations the chances that a mutation will be established depend in greater part upon the absolute number of times of occurrence, which in turn depends upon rate of mutation and population size. From Fisher's work (1930) it appears, for instance, that a mutation of definite selective value (.01) arising at the rate .000001 in a population of 10,000,000 is sure to become established within 25 generations, but in a population of 10,000 it may require 25,000 generations—a time so long that in many cases the selective value or other limiting factors are likely to change.

Random loss of allelomorphs from a population tends to be at the rate of about $\frac{1}{2} N$ per generation in populations with equal numbers of breeding males and females and under other breeding conditions may become $\frac{1}{4} N$ or smaller (Wright). In large populations this effect is negligible and is almost certain to be offset by repeated mutation, but in very small populations the effect may be important. In large populations the chances of fixation of a mutation depend mainly on selective value, but in small populations the result is largely random and independent of selection. An adverse dominant has virtually no chance of fixation in a large population, but this chance may approach $\frac{1}{2} N$ and be appreciable in a small population. As regards relationship to selection, Wright shows that there is a critical population size at about $N = \frac{1}{2} s$ (in which s is the coefficient of selection as in Fisher and Wright); above that size selection is the important modifier of gene ratios, while below it random effects may outweigh selection. This value is small in comparison with natural breeding populations; $s = .001$ is slight selection pressure, and even at this low level the critical value of N is only 500, and for $s = .01$, $N = 50$. Most breeding populations are larger than this. Random effects will, however, also occur at population levels well above the critical value, but they will be less frequent and only rarely of significance for continued evolution.

In all cases genetic intragroup variability tends to be proportional to the size of the breeding population, although again there are critical points above which little further increase in variability occurs (Wright). If selection rate is low relative to mutation rate, this value is about $¼ u$ ($u =$ mutation rate). For instance if $u = .0001$, increase of breeding population above about 2,500 will produce little increase in variability, provided that s is less than .0001; but this is a relatively high mutation rate and a very low selection rate. In the more usual circumstances, when s is greater than u, the critical value is about $N = ¼ s$. With a small selection value of .001, this critical point is $N = 250$, a small population. Apparently the range within which the control of variability by N is significant is low under natural conditions and this effect is less important in most instances in wild populations than its emphasis by Wright might imply.

Wright suggests that very small populations will have low variability. Most of their genes are fixed, selection has little effect, and transfer from one fixation to another, depending on mutation rates and random elimination, is slow. Hence evolution would tend to be slow and inadaptive and to tend toward extinction. On the other hand, very large populations may be highly variable and selection is effective and rapid. Fisher and, with some modification, Haldane consider this as the optimum condition for rapid evolution, but Wright points out that this is true only within the limits of existing variation and that the building up of further variability will be very slow, since strong selection tends to eliminate variability in large populations and weak selection tends to end in a static condition of fixed gene ratios. Large variable populations favor rapid racial differentiation, but, as has been noted in discussing variability, such conditions are not propitious for sustained rapid evolution.

Wright concludes that the conditions most favorable for evolution are found in populations of intermediate size—large enough to promote variability and to make selection effective, small enough to have random fluctuations of gene frequencies. This favorable size cannot be well defined in its own terms, but is estimated by Wright to be approximately at a point where u is not much greater than $\frac{1}{4} N$ and s not much greater than u as an average for the genes concerned. With different values of these other variables, such a population of medium size might usually be between 250 and 25,000.

Within the limitations set and for sustained progressive evolution this conclusion seems probable, although it has not yet been subjected to adequate experimental or observational checking. Another problem is, however, of much importance for the study of special aspects of evolution: the influence of population size on maximum rates of evolution, regardless of direction and with the postulate that other factors tend toward maximum rates. Such maximum rates would depend upon the average number of generations required to eliminate a gene (everything, including chance, favoring such elimination) or to make its frequency 100 percent (everything favoring its spread). This number of generations and therefore these maximum rates vary inversely with the size of the breeding population. Thus the most rapid possible evolution must occur in populations of minimum size. Such evolution will usually lead to extinction, but in a large number of trials it will not always do so.

To this conclusion as to very rapid evolution in very small populations the objection has been stated that the absolute incidence of mutations in such populations would be too small under natural conditions to provide sufficient materials for evolution. Although I have not attempted adequate mathematical analysis (which is beyond my powers), I cannot see that this objection is decisive. The utilization of mutations in small populations is more efficient, that is, a single mutation has a much greater chance to survive or to become universal in the population and can do so much more rapidly. This may largely compensate for the lower absolute mutation rate. With an average rate of .00001, which is probably conservative for small spontaneous mutations, each active locus would produce an average of one mutation in 200 generations in a population of 500—a low absolute rate, indeed, and quite insignificant in a large population, in which such isolated mutations would have virtually no chance of becoming characteristic of the group. But in so small a population there is such a chance, and during 1,000,000 years each of the active loci would average 5,000 new mutations with one-year generations or 500 with relatively long ten-year generations. Thus, under relaxed selection inherent in the situation it is likely that such a population, if it survived for such a period, would have changed its entire genetic constitution, as regards the active loci, at least once and probably several times. It would almost certainly become a new genus and might well initiate a new family or a still higher group. In larger populations dominated by selection, a million years may not suffice to introduce a new species and relatively seldom suffices for evolution of generic rank. Note the previous estimate that the actively and progressively evolving tertiary horses took on an average 8,000,000 years for evolution from one genus to the next.

Other situations, more special, but of frequent natural occurrence, may provide small populations with greater variability and more extensive evolutionary materials than could arise from their own absolute mutation rates. In a model stressed by Elton (1924) the population fluctuates greatly in size. A good summary of such a cyclic variation is given in Hamilton (1939). The incidence of mutations will be in proportion to the average population, while the breeding population effective as regards the subsequent fate of these mutations will be a smaller number, near the minimum population size. Haldane (1932) has criticized Elton's conclusion (congruent with, but different from, that here reached) that random gene extinction may be important in such cases by showing that in the Arctic fox of Kamchatka, varying approximately from 800,000 to 80,000 every three years, a gene would have an even chance of extinction only after 330,000 years. In reply it may, however, be pointed out (a) that this is not really a long period as species go, (b) that much greater fluctuations with lower minima occur (e.g., in varying hares, see MacLulich 1937), and (c) that probability of random extinction less than .5 may strongly influence evolution without dominating it. In any case, the fact remains that in such a group the effective mutating population is considerably larger than the effective breeding population.

This whole subject of animal population cycles is one that may prove to be crucial for evolutionary studies, but its value is still potential, because practically everything about these cycles is now in dispute except the fact that they do occur. One of the possibilities is that the cycles may prove to be correlated with some radiation phenomena, for instance those accompanying sunspot cycles (see Hatfield 1940). This possibility has enhanced interest because of the further possibility that some forms of natural radiation affect mutation rates, as X-rays are known to do in the laboratory. Population cycles might thus be accompanied by mutation cycles, and the combination would inevitably have an effect on both the rate and the direction of evolution. At present, however, every element in this possible concatenation is an almost completely unsupported hypothesis, which is mentioned only as a suggestion worthy of investigation.

Wright stressed a model that is of great importance both in this connection and in other connections, that is, a large population differentiating into local groups. When the latter are small, the fate of the variability of the large population may be quickly determined by more or less random effects within the small populations. This process is not continuous and is almost immediately self-limited; but if the groups are interfertile and there is a limited amount of migration or interchange between them, each may continuously have what amounts to an incidence of mutation larger than could occur in the small effective breeding population.

Finally, the coincidence of increase in mutation rate with decrease in breeding population would effectively provide for great acceleration in rate of evolution. That such coincidences actually occur cannot be disproved, but they must be rare, and the hypothesis does not seem necessary to explain known facts. Nothing is really known as to fluctuations of mutation rates in nature, and there is no apparent reason for a correlation, direct or indirect, with population size. The chances of such coincidence cannot be calculated from present data, but will probably prove to be small.

In summary, very large populations may differentiate rapidly, but their sustained evolution will be at moderate or slow rates and will be mainly adaptive. Populations of intermediate size provide the best conditions for sustained progressive and branching evolution, adaptive in its main lines, but accompanied by inadaptive fluctuations, especially in characters of little selective importance. Small populations will be virtually incapable of differentiation or branching and will often be dominated by random inadaptive trends and peculiarly liable to extinction, but will be capable of the most rapid evolution as long as this is not cut short by extinction.

Estimates of actual population sizes are difficult and subject to large errors, but much attention has recently been paid to such estimates for recent animals, especially birds and mammals (e.g., Blair 1941), and a large number of relatively reliable census estimates are available. Less attention has been paid directly to the problem of effective breeding populations, but here, too, many estimates await synthesis, and data are available in many groups for approximate conversion of census to breeding estimates.

As regards fossils, very few numerical estimates have been attempted, and these are so uncertain that the use of numbers is more likely to be misleading than helpful. If only as a curiosity, an example may be derived from Broom's discussions (1932) of fossil reptiles in the Karroo. He estimates that there are remains of about 800,000,000,000 animals preserved in this formation over an area of about 200,000 square miles, and he considers the time involved to be at least 40,000,000 years. As an average, about 20,000 animals per year are thus supposed to have been buried and preserved as fossils. Going beyond Broom, it might be estimated, with even greater uncertainty, that about 1 in 5,000 of the animals living in this area in any one year died and was preserved as a fossil. Then there were 100,000,000 of these reptiles living over the 200,000 square miles at one time on an average—a population density of about 500 per square mile for all species (most species were large in comparison with modern reptiles, but many were comparable to living lizards in size). Another, and still more uncertain, estimate is that the number of distinct breeding groups in the area at any one time was on the order of 1,000, each of which would thus include about 100,000 individuals. In each group about ½ may have been actively breeding in any one year, giving an average $N = 50,000$ for the species found as fossils in the Karroo. The estimate is not unreasonable, but the true figure may have been 5,000 or 500,000, as an average, and individual lines probably varied enormously, not improbably from as low as 50 to as high as 5,000,000 at various times.

Although such estimations are really futile, it is often possible to infer in a qualitative way whether fossil populations were more probably large or small. The enormous variations of fossil sampling under the influence of chance and many other factors may make valid inference impossible, especially the negative influence of small populations, but a reasonable conclusion may frequently be reached. As regards single species or genera at one locality or a few localities, the negative inference is never warranted. By chance alone a large population may leave very few recoverable remains in a small number of deposits in which it is, nevertheless, represented. For instance, late Pliocene horses of the subgenus *Equus* (*Plesippus*) were very rare in collections until 1928, when a deposit was discovered near Hagerman, Idaho, that contained remains of hundreds of individuals of this group. On the other hand, a species with consistently very small populations can leave no such an accumulation unless under exceptional and usually recognizable conditions that result in the concentration of the remains of a long period of time. Thus, in the above example, as in analogous cases, no direct inference as to population size was permissible until the discovery of the rich deposit, which warranted the inference that the population size was large, at least in some localities.

Surer inferences, both positive and negative, can be based on larger and more widespread series of occurrences within a phylum. If an animal group is continuously present in areas of deposition suitable for its subsequent recovery in fossil form and if the average population is large, single occurrences may be found; but among many different discoveries, one or more is very likely to include numerous individuals. On the other hand, under the same conditions a phylum with consistently small average populations will eventually leave a sequence of remains; but all occurrences will consist of one or a few individuals. Other factors must also be considered, especially the chances of fossilization, which vary greatly so that an abundant group with small chance of fossilization can leave a record similar to that of a rare group with good chance of fossilization. With reservations introduced by such effects, it is often a valid inference that a sequence of isolated occurrences represents a group with small populations and that an equal sequence of more numerous occurrences, some of which are abundant, represents a group with large populations, the inference being more probable the longer the sequence. Representative data of both sorts are exemplified by the American occurrence of Apatemyidae (an unusual extinct family of rodent-like primates) and early Equidae over roughly equivalent spans.

TABLE 14

OCCURRENCES IN AMERICA OF APATEMYIDAE AND EARLY EQUIDAE

	APATEMYIDAE		EQUIDAE	
Age	Known Specimens	Localities	Specimens in American Museum	Localities
Middle Paleocene	1	1	Unknown	
Upper Paleocene	5	2	Unknown	
Lower Eocene	1	1	397	Many
Middle Eocene	7	2	54	Several, 2 fields
Upper Eocene	1	1	11	Few, 1 field
Lower Oligocene	1	1	30	Several
Middle Oligocene	Unknown		125	Several
Upper Oligocene	Unknown		39	Several
Maximum known from one area and stratum	4		About 200-300	
Average individuals per occurrence	2		About 20-40	

The remarkably continuous record of the Apatemyidae, with very few individuals of each age, strongly suggests intensive sampling of consistently small populations. Although probably less chance of fossilization and recovery is a contributing factor, if it were the dominating factor and the populations had actually been large, the record would probably be more spotty and variable. The almost fully continuous record of early horses, with great variation in number of individuals recovered as fossils from the various occurrences, and some occurrences very abundant, almost surely represents sampling from large populations.

A record like that of the Apatemyidae, which is exceptionally good for rare fossils, gives a fair idea of actual structural change in the phylum, but provides almost no direct evidence as to variation and other essential factors underlying this morphological evolution. Most of the principles and theories of evolution advanced by paleontologists have been based upon groups with long, continuous, and relatively abundant records; hence, on groups that probably had large breeding populations. Important theoretical studies in this field involve, for instance, the abundant fossils of trilobites, graptolites, brachiopods, gastropods, ammonites, oreodonts, titanotheres, and horses and within these the particular phyla and stages that are most abundant.

The recent work on relationship of population size to evolution shows that principles so based cannot reasonably be accepted as generalizations. As far as these principles are logically derived from adequate data, they are valid descriptions of evolutionary processes in particular groups at certain times, but as far as these facts can show, the processes may have been quite different or have had quite different results in the more numerous and in some respects more important groups that are poorly recorded as fossils.

Recognition of this limitation does not make paleontological data less useful in the study of evolutionary theory. On the contrary, it permits better co-ordination of paleontologically and experimentally developed theories, which have often been radically discrepant. It gives a better basis both for the checking of neobiological observations against the fossil record and for the interpretation of that record in harmony with what is known of the genetics and other physiological processes of living animals.

SELECTION

The role of selection.—No theorist, however radically non-Darwinian, has denied the fact that natural selection has some effect on evolution. An organism must be viable in an available environment in order to reproduce, and selection inevitably eliminates at least the most grossly inadaptive types of structure. Aside from this obvious fact, theories as to the role and importance of selection range from belief that it has only this broadly limiting effect to belief that it is the only really essential factor in evolution.

Attacks on selection as an important evolutionary factor have been based on diametrically opposite considerations. Many real evolutionary occurrences have been, or have been believed to be, nonadaptive and hence inexplicable if selection is a controlling influence. On the other hand, various evolutionary phenomena have been considered so minutely adaptive that it was asserted that selection cannot have been efficient enough to produce them. Thus, it may be argued that there is no conceivable selective advantage of one precise ammonite suture pattern over another of equal complexity and that the evolutionary process may even result in structures that are positively disadvantageous—the antlers of the extinct Irish stag supposed to be a classic example. Again, it is asserted that structures which are probably eventually of selective value, such as titanothere horns, arise gradually as rudiments that apparently had, at first, no use and therefore no selective advantage. Similarly, mimicry has been supposed to be of no advantage until well developed and also to be carried to extremes well beyond the point where greater perfection would give greater selective value.

The weight of such objections has led to a series of alternative theories that have in common only the minimizing of the effects of selection. Three extremes are especially outstanding among such theories. The first, antedating Darwin (and not entirely rejected by him, despite his emphasis upon selection) was the Lamarckian theory that adaptive structures are not selected, but are caused by environmental influences and by individual efforts to meet the exigencies of life. Although open to some of the same objections as is selection, this provides a logical alternative explanation of adaptation. Particularly appealing to paleontologists, neo-Lamarckian theories reached considerable subtlety and complexity at their hands (e.g., Cope) and seemed at one time, and to some thoroughly competent students, sufficient to explain all the essential morphological observable facts. Experiments in heredity in the present century, however, not only have failed to corroborate that there is such a process but also have shown that it is highly improbable, if not impossible. I do not propose to consider the theory further in the present study. It is, perhaps, out of the question to offer a rigid proof that Lamarckian evolution could not occur and never did occur, but it is neither a necessary nor a sufficient hypothesis to explain the pertinent facts of evolution as seen by the paleontologist.

Another extreme is seen in various theories that suppose adaptive characters to have arisen, not in response to external influences and individual needs or at random with respect to them, but in anticipation of them. These theories avoid the necessity of providing a mechanism, because of which the neo-Lamarckians broke down, by begging the question. Concluding that the controlling factor of adaptation and of evolution in general is nonmechanical, they name this factor ("entelechy," "aristogenesis," and so forth), but no explanation is provided, and the definitions of their terms say little more than that they designate unknown causes of known phenomena. This appeal to the unknown, inherent in all such theories, however obscured by the naming fallacy, is metaphysical, not scientific. The categorical rejection of all physical hypotheses does not warrant the proposal of a metaphysical hypothesis as a supposed part of scientific interpretation. Recourse to such devices, impelling as is their fascination, has always resulted in stultification. The metaphysical explanation might, indeed, be true, but no hypothesis for which there can be no rigidly objective test should be accepted as a conclusion or working principle in research.[13] The greatest vogue for such theories has now passed, although they always have adherents and may well come into fashion again. Emphasis upon spiritual values, proper and desirable in times of stress like the present, sometimes has an undesirable and unnecessary concomitant of depreciation of scientific values, even among scientists.

The main objections to Lamarckian explanations (which lack any known mechanism) and vitalistic explanations (which deny all mechanism) are successfully met by a third extreme that proves, nevertheless, to have objections of its own. This theory, usually called "preadaptation" and associated with the name of Cuénot, who emphasized and expounded it in greatest detail (1921, 1925), maintains, in essence, that new types of organisms arise at random by Mendelian mutation and that the types that survive are those that happen by chance to be viable in an environment that they are able to find. This theory depends only upon demonstrated mechanisms, mutations and gene inheritance, which are primary facts, however unexplained they may themselves be. It is consistent with the data of experimental genetics, as the extreme alternative theories were not, and for a time it was accepted among experimentalists as the only theory consistent with their data. It was, however, almost universally rejected by paleontologists and by many observational neozoologists and was one of

[13] As a matter of personal philosophy, I do not here mean to endorse an entirely mechanistic or materialistic view of the life processes. I suspect that there is a great deal in the universe that never will be explained in such terms and much that may be inexplicable on a purely physical plane. But scientific history conclusively demonstrates that the progress of knowledge rigidly requires that no nonphysical postulate ever be admitted in connection with the study of physical phenomena. We do not know what is and what is not explicable in physical terms, and the researcher who is seeking explanations must seek physical explanations only, or the two kinds can never be disentangled. Personal opinion is free in the field where this search has so far failed, but this is no proper guide in the search and no part of science.

the reasons for the rift between experimental and observational students of evolution. The picture of evolution offered by this theory, in its purest form, is essentially random, while the picture seen by paleontologists was essentially orderly. Not all, surely, but many phyletic lines progress in a regular manner toward a condition that has appeared to some students so fixed as to be predestined—a condition that certainly was not the only one possible for the phylum. If, then, the direction was determined by the nature of the mutations alone, and limited only by a viable relationship between organism and environment, it seems impossible that the mutation should be random, and one is driven back to the idea of purely directional mutation, which seemed physically inexplicable and therefore encouraged metaphysical speculation.

The last word will never be said, but all these disagreements can be reconciled, and the major discrepancies can be explained. In the present synthesis adaptation, preadaptation, and nonadaptation all are involved, and all can be assigned immediate, if not ultimate, causes. That all basically depend upon mutation is merely a matter of definition. That the mutations are spontaneous and random, at least in the special sense elsewhere defined, is a conclusion warranted and, with some restrictions, demanded by the experimental data. Calling them "spontaneous" means simply that their causes are unknown and that they are not orderly in origin according to the demands of any one of the discarded theories. The incidence in time and the individuals affected seem to be random or nearly so in the same sense—that they do not agree with hypotheses that assume a more specific incidence. That the direction of mutations is entirely random is certainly not true; but neither is it true that mutations regularly occur in one direction only. Given a certain hereditary type of developmental pattern, the changes that can occur in it and their effects upon the structures developed are strictly limited, and alternative changes are not introduced in exactly equal numbers; but in almost every case the change can be and is in at least two, frequently more, different directions.

Many of these directions, most of them if conditions are stable, are nonadaptive. As long as these changes are neutral with regard to selection, they present no particular problem. The number of changes that are really completely neutral is probably smaller than the critics of natural selection grant, but without doubt they do occur. Robson and

Richards (1936), have gathered examples, some but not all of which seem to be of this nature. Some students hold that the differences between genera and higher groups may be adaptive or have selection value, but not the differences between subspecies or species (e.g., Jacot 1932); others use similar data to attack the same theory by the diametrically opposed conclusion that local differences between subspecies are adaptive, but that generic distinctions are not (e.g., Goldschmidt 1940). Some particular objections of this sort have been met by demonstrating that characters hitherto supposed to be nonadaptive are quite definitely adaptive (e.g., Dunn 1935), and it has also been proved beyond much doubt that characters of unknown adaptive value are nevertheless definitely selected by differential survival (e.g., Dunn 1942). In the nature of things it is quite impossible to establish that every single genetic difference between two populations has selective value, and probably some distinctions differ in this respect; but neither is it possible to prove that any are really indifferent, and this is certainly untrue of many and probably untrue of most. See also the discussion of small progressive steps in trend evolution, Chapter V.

In the preceding section there were summarized studies by Wright that show that in small populations selection is relatively ineffective in controlling the spread of gene frequencies. Under these conditions even disadvantageous mutations may spread rapidly and come to characterize a whole population, as long as they do not reach the rigid limit of complete inviability. Such random effects amply allow for the occasional rise of inadaptive organisms, which would, however, usually become extinct within a relatively short time when the disadvantageous genes accumulated. Theoretically such occurrences should be rare, and in fact they seem to be so in large populations. Thus, Osborn (e.g., 1927), who drew his conclusions from an enormous body of data relating almost entirely to large populations, was able to believe that "speciation"—hence all evolution, which was for Osborn a multiple of speciation—"is apparently always adaptive."

Next, it has become more clearly realized that the result of selection is to increase or decrease the proportions of certain specific genes or genetic combinations and arrangements in populations, but that it does not act directly on the genetic units. It acts on structures developed under the influence of the hereditary factors, but without a point correspondence with the latter. Elimination of a disadvantageous structural character can thus mean the necessary and simultaneous elimination of an advantageous structure, because the two are not separable or are incompletely separable genetically. Conversely, selection for an advantageous character may necessarily develop a disadvantageous character. Correlation in characters does not *ipso facto* reveal this necessary genetic connection—a point stressed by Robson and Richards (1936), who give examples of what I believe to be true correlation, as well as of false genetic correlation—but it certainly exists, and the fact that the necessary experimental check is frequently impossible does not invalidate the use of this hypothesis to explain apparent evolution contrary to selection. Thus, horse skull proportions and gross size (Robb 1935a) are two characters apparently with a single genetic control, and the chances for the selection of the genetic determinant are increased because one direction of change is advantageous in both characters. In the case of ear length and palatilar length in the mouse, ear length is certainly subject to selection (e.g., for shorter ears in colder climates), and it involves coincident skull changes that can be either selectively neutral or disadvantageous. In cases such as that of the Irish stag (see later discussion of momentum), one character may be carried to a point where it is disadvantageous because it is genetically linked with a character that has not yet reached its optimum. The effect of selection upon a gene or genetic system is not the effect upon any one character, but a resultant for all the characters that are affected by any one hereditary unit.

The same sort of correlation may induce the rise of structures that have little or no adaptive value in incipient stages, but that later become adaptively significant and so re-enforce and accelerate the process. Hersh (1934) suggests that this is true of the titanothere horn, an example that Osborn (e.g., 1929) used as evidence for his idea of germinal predisposition. This example is not absolutely convincing, but it is suggestive, and the process is inherently probable, whether or not it can be completely demonstrated as the sole possible explanation in any one case.

In another respect this is a special case of preadaptation. If there are demonstrated mechanisms by which characters may arise either without regard for or opposite to the pressure of selection, it must sometimes happen that the characters are actually advantageous in some other available environment (in the broadest sense, or mode of life, ecological niche, and so forth). The essential is that the change is inadaptive with respect to the ancestral group and its way of life, but that once it has occurred it becomes adaptive for some other conditions that are at hand or do subsequently arise. The first phase is preadaptive and not controlled by selection, but it leads nowhere unless in a second phase it becomes adaptive and is spread and continued by selection. Thus, among small populations, with their random changes in gene frequency, nine hundred and ninety-nine might be merely inadaptive and might lead to early extinction, as Wright concluded, but the thousandth might be preadaptive. Relatively rare as such cases must be, they are potentially and, as will be shown to be probable in the sequel, really of tremendous importance, because they afford an explanation of quick, radical shifts in adaptive types.

It has repeatedly been emphasized and, with somewhat dubious logic, urged as disproof of the importance of selection that its role is purely negative, that only genetic mutation creates, while selection prunes and chooses among the indefinitely large supply of creations. If this were entirely true, selection would still be a most essential factor in evolution, but it might be proper to rule out adaptation in nature except in the qualified sense of preadaptation. Only mutation supplies the materials of creation (whatever a mutation may be or however it may arise), but in the theories of population genetics it is selection that is truly creative, building new organisms with these materials. In most natural populations the number of possible combinations of discrete hereditary units into individual zygotes is enormous, and only a small proportion of them can be realized. To the extent to which selection controls the frequencies of the different units in the population, it determines, within limits, which combinations will be realized and in what proportions (Fisher 1930). It does not, then, simply kill off or permit to live fixed types of organisms delivered to it by the random action of the laws of heredity, as seems to be inherent in the cruder forms of the preadaptation theory of evolution. Selection also determines which among the millions of possible types of organisms will actually arise, and it is therefore a truly creative factor in evolution.

Intensity of selection.—Selection is a vector that has both direction and intensity. A priori, the effect of varying intensity of selection is simple enough. Increasing intensity should accelerate adaptation and hence evolution in a group not already perfectly adapted, and at the same time it should decrease variability by eliminating deviation from the optimum. In nature this does appear to be one basic tendency of selection, but the actual situation is more complex.

The determination of intensity of selection is in itself a problem to which there is apparently no direct approach and one which is very difficult to treat practically. An indirect measure has been used in various forms in the studies of Fisher, Wright, and Haldane, discussed in the preceding section. Without attempting a summary of their difficult mathematical treatment, it may be said that in various ways they all estimate the intensity of selection by the pragmatic test of the rate of increase in the proportions of a given gene from one generation to another. In the simplest possible situation a coefficient s might be so defined that with equal chances of occurrence selection would cause a gene, A, to be present in a following generation in n individuals, while its allele, a, would be present in $(1-s)n$ individuals; s is then the coefficient favoring A. In less simple and more natural situations the calculation is more complicated, but the general significance is the same. Complete selection, $s = 1.00$, is known in cases where the alternative is wholly lethal. Other coefficients, down to about .01, have been determined with reasonable certainty. "It is," according to Wright (1940), "probable that most of the mutations which are important in evolution have much smaller selection-coefficients than it is practicable to demonstrate in the laboratory."

Although the three students here cited (Fisher, Wright, and Haldane) differ sharply on some points, all agree that evolution without selection would be extremely slow in large populations, in which such events as the spread of a single gene through a population might take periods covering as many generations as there are individuals in the population if controlled by mutation and random elimination alone. Even a subspecific advance in the Equidae might well have taken approximately 1,000,000 to 10,000,000 years if determined by factors without natural selection. Since the whole sequence from *Hyracotherium* to *Equus* actually occurred through nine genera in about 45,000,000 years, the process was much faster than mutation alone would seem to permit. There must have been some other factor, and all the evidence suggests that this factor was natural selection.

These students also agree that surprisingly slight selective advantages may control the direction of evolution and may have strong effects in accelerating its rate, thus answering the objection that many characters developed under the supposed influence of selection have selective advantages too slight to have any effect. It is true that the answer is theoretical, but the objections are even farther removed from objective observation, being based merely on unchecked hypothesis or unreliably subjective opinions as to probability. Haldane states that a gene cannot be considered neutral (i.e., selection does not fail to have a significant effect on the gene frequencies) unless sN is on the order of 1 or less, and Wright makes the more exact, but sufficiently concordant, statement that the critical value is about at $Ns = .5$. For so low a selective advantage as .0001, lower than has been conclusively demonstrated for any mutation on which an actual test has been made, selection would still be the dominant factor as regards the fate of the gene in a population of 5,000 or more breeding individuals. In smaller populations or for still smaller coefficients of selection, random effects and mutation rates may become dominant, as has already been sufficiently emphasized.

If selection favors a single fixed genetic combination, Wright has shown that it would have little effect in decreasing variability unless $s = 1/8N$ or more. Stronger selection will greatly reduce variability, until it will be almost eliminated at $s = 4/N$. For a population of 100,000 the value of s is only .00004, a value so low that it really proves too much. According to this characters with appreciable selection value would be almost invariable in large populations, whereas in fact adaptive characters in such populations do always have a considerable amount of variability, both genetic and morphological. This is, however, readily accounted for by several factors making for variability in natural populations and not involved in this model. Mutation rates always provide some variability that cannot be wholly eliminated by the strongest selection. In a large population, spread over a considerable area, the optimum will generally tend not to be exactly the same at all points, so that even strong selection does not tend toward one point and the same sort of effect is seen in groups subject to cyclic or secular fluctuations of environment and hence similarly without one permanent optimum. Under such conditions there is an optimum amount of variability that is itself maintained by selection, as has been previously discussed. Within the common pattern of a large species divided into semi-isolated, inter-fertile breeding groups variation will also be maintained in the face of the strongest selection—intergroup variation by the differences in optima for the various groups, and intragroup variation by migration and marginal interbreeding.

In spite of the paucity of adequate observational data it is clear that selection does, indeed, influence variability. In a previous section it was shown that functional, integrated structures of adaptive significance—i.e., those that are under strong selection pressure—tend to have low, although still appreciable, variability and that the variation differs surprisingly little from time to time for one variate within a group, for homologous variates in different groups, and for different analogous variates. On the other hand, vestigial and nonfunctional characters—i.e., those with low selection value—have higher average variability.

The influence of intensity of selection on rate of evolution is not so directly demonstrable. Strong selection may favor rapid evolution or may bring evolution to a dead stop, depending upon the direction of the selection. Weak selection simply transfers determination of rates to relatively random influences, the effects of which depend chiefly upon the size of the population, producing slower evolution in large populations and faster evolution in small populations.

Direction of selection.—The characteristics of selection as a vector are intensity and direction. Direction may be centripetal, centrifugal, linear, or a resultant of combinations of these (Fig. 10). The direction is defined by the relationship of the morphology and the physiology of a population to the environmental field in which the organisms function. If the modal condition of a character in the population, that "typical" for it, has continuously greater survival (or, more strictly, reproductive) value under the given conditions than any variant or newly mutant condition, the population is well adapted as regards that character. Selection is then entirely centripetal. It acts to eliminate the variant forms, and it tends to reduce variation in any direction and to concentrate the population around a point, the optimum condition. This involves an increase of specialization, not necessarily in the usual sense of progressive adaptation of all individuals, because some are already postulated as perfectly adapted, but in the sense of reducing the number of less well-adapted individuals in a population as a whole.

This is an evolutionary change, but in the more usual sense there is no evolutionary advance, because the mean characters of the population do not change. Centripetal selection is, in fact, directed against any such shifts in means.

If, on the other hand, the modal condition of a character in a population is so ill-adapted that any variant condition is more advan-

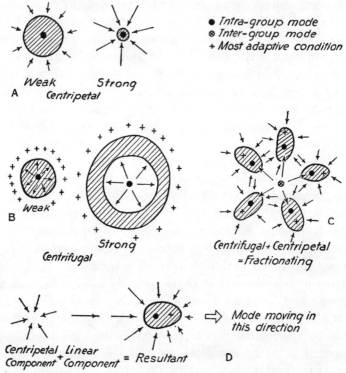

Fig. 10.—Selection vectors. Heads of arrows show direction of selection, and lengths of shafts are proportional to intensity of selection. Shaded areas show range of variation covered by population.

tageous, selection is centrifugal. It favors increase of variation and progressive divergence from the population mean. As will presently be discussed, this selection can rarely be exactly realized in nature because of the other factors involved as a model, but it will serve for a theoretical analysis of the single factors underlying more complex real phenomena. In the simplest model, strong centrifugal selection would eventually result in the elimination of average individuals from the population, so that its distribution in the morphological field would be an expanding ring or a hollow circular band. The average for the character in question would remain the same, but there would be no individuals actually at the mean in this respect. Even as a model, this pattern is obviously too simple; it is made more useful if the inevitable centripetal forces within the parts of the ring are also taken into consideration. Then the ring breaks up into smaller groups, in each of which selection is actually linear, away from the intergroup mean, and the situation is evidently less abstract and closer to reality than when the unoccupied central point is considered an intragroup mean.

Pure centripetal selection occurs in nature, but it is doubtful whether it usually endures very long. Pure centrifugal selection probably does not occur in nature, although it is surely a component of more complex selective forces in many cases.

Centripetal selection can remain unmodified only as long as the optimum is completely static and selection is continuously strong enough and effective enough to prevent any considerable deviation of population mean from the optimum. In some unusual cases (see "Low-Rate Lines" on p. 125) this delicate equilibrium has apparently been maintained even for millions of years, but in the great majority of cases it is continuously or continually disturbed. The simplest, or most easily understood, disturbance is some change in the external environment that causes a slow shift in the optimum. Although still centripetal, bearing toward the center from all directions and tending to concentrate the population on the optimum, selection then becomes stronger on one side than on the other, favoring variants on the side toward which the optimum is moving, and it acquires a linear component superposed upon the centripetal pattern. Progressive shift in the mean occurs, and as long as selection is effective this evolution will occur as nearly as possible at the same rate as the shift in the optimum. Any considerable lag is likely to leave the population behind the moving selection field and subject to a different selection pattern, resulting either in extinction or in a new direction of evolution.

In the broadest sense, environment is so complex and subtle that the optimum may shift in what appears at first sight to be a stable environment. Thus, in animals capable of self-directed locomotion the optimum for size shows a tendency to shift slowly in one direction, toward larger size, a tendency that may be counteracted by other influences and abruptly reversed, but is nevertheless usual in the evolution of such groups. Even though individual animals may be perfectly adapted at a particular size level, in the population as a whole there is a constant tendency to favor a size slightly above the mean. The slightly larger animals have a very small but in the long run, in large populations, decisive advantage in competition for food and for reproductive opportunities and in escaping enemies. An abrupt increase in size in a small number of individuals at any one time is more likely to be disadvantageous, even if genetically possible, because it may not be accompanied by other necessary independent adaptations to a larger size, correlated genetic changes may be inadaptive, and abnormally large individuals may be subjected to the hostility of competitors and potential mates. Thus, populations that are regularly evolving in this way are always well adapted as regards size in the sense that the optimum is continuously included in their normal range of variation, but a constant asymmetry in the centripetal selection favors a slow upward shift in the mean.

This is, I believe, the causal background of the empirical paleontological principle that most phyla have a steady trend toward larger size. As regards abundant groups in which selection is continuously dominant as an evolutionary factor this generalization has few exceptions —so few that it is commonplace in the literature of vertebrate paleontology to read that a certain species "cannot" be ancestral to another of later age because the earlier species was larger. There are, nevertheless, a good many exceptions that are virtually certain and more that are probable. These can invariably be explained either by special circumstances that make selection favor small size (e.g., very warm climate for homothermous animals, insular conditions, subterranean life, limitation of food supply, greater opportunities for escape from enemies because of smallness), or by other circumstances (such as small population size) that relax the control of selection.

A change in one character may also change the optimum of another even though they are genetically independent and seem to be independent as regards adaptation. Thus, an oscillating shift of both optima may operate continually and in a single direction for each. For instance, both brachyodont and hypsodont teeth occur in large and in small mammals, and hypsodonty and gross size are genetically independent and, as regards extremes, adaptively independent. Nevertheless, nominally brachyodont large animals almost invariably have somewhat higher-crowned teeth than their likewise brachyodont smaller allies or ancestors. The comparison of the large brachyodont *Hypohippus* with its small brachyodont ancestor *Hyracotherium* (data in Chapter I) is a typical example. Even if no significant change in diet is involved, larger animals live longer, on an average, and require more food. They need teeth having not only larger working area but also more durability. We, ourselves, although brachyodont, have relatively higher tooth crowns than do most small primates of similar omnivorous food habits. Such secondary adaptations may limit the primary adaptation. An increase in size disproportionate to the increase in durability of teeth is disadvantageous. When tooth durability has caught up, increase in size again becomes selectively valuable, making more increase in durability advantageous, and so forth, in a potentially endless cycle, while all the time the external environment may be entirely static.

Still more marked but less frequent changes in optima may occur when a mutation that was inadaptive in its original genetic surroundings happens to survive and to become fixed. Then the optima are likely to shift for all other genes the expression of which has any direct or indirect functional relationship to that of the mutation. Thus, pleiotropic genes, polygenic characters, and adaptively correlated characters form a mesh so intricate that a single change at any point may initiate a series of reactions by which selection eventually produces a change in the whole system.

The most generalized abstract model of centrifugal selection can hardly be related to a closely similar example, because it is difficult (I find it impossible) to imagine a situation in which any change whatever in a character would be advantageous and could physically occur in an unlimited number of different directions. There are, however, real and important instances in which more than one of the limited number of variant types in a population would be better adapted to available environments than the modal types, and this is a special case of centrifugal selection. Essentially such a process, one of the most common forms of evolution on low taxonomic levels, occurs when a variable and relatively unspecialized population divides into groups, each of which is adapted to a different ecologic niche. In longer-range phyletic studies a manifestation of similar centrifugal selection is commonly seen when branching is accompanied by evolution in two or more opposed directions from the ancestral type. Thus, the shell of shelled animals impedes motion and requires the utilization of food and vital processes that are unnecessary for reproduction and possibly detract from it. In regard to these factors thin shells or no shells will be favored by selection. On the other hand, the shell is a protection from predators and other unfavorable environmental conditions, and in regard to this factor heavier shells are favored by selection. Aside from such common developments as the strengthening of shells by ribbing instead of by thickening, the character generally evolves only in one of two directions—more shell or less shell. Under varying balances of the opposing selection factors, almost every intergrade develops; but very thin shells, not otherwise strengthened, are rare. They do not suffice for adequate protection, yet do not fully realize the advantage of maximum reduction. Here there is a balance point where variation in either direction is an advantage over the mean condition—centrifugal selection—and in fact among otherwise nearly allied mollusks some forms may become shell-less and some relatively heavy-shelled. Thus, *Chiton*, strongly shelled, and *Neomenia*, unshelled, seem to have had a common ancestry at no very remote time, and allied shelled and shell-less forms (such as slugs and land snails) occur in several different groups of gastropods.[14]

Other examples are provided by the morphologically simple character of size, which can vary only in two directions, either of which may, under different conditions, have greater selective value than an intermediate condition. This sort of dichotomy occurred at least twice in the history of the horse—when late *Miohippus* split into one line smaller than the ancestry, *Archaeohippus*, and another line larger, *Parahippus*, and one division of *Merychippus* similarly split into small *Nannippus* and large *Hipparion*.

If, through relaxation of selection or some other cause, the variability of a group increases and the specificity of its adaptation decreases, several or many of the large number of different genotypes are likely to have some advantage in particular environmental conditions, and these may be driven away from the ancestral modal condition (even if part of the population remains there and is sufficiently

[14] As regards this particular example there is an alternative theory that the shell-less forms are primitively so, never having had a shell. In some cases this may be true, but it seems almost certain that some, at least, represent a process of loss of shell in one branch of a phylum while the shell became stronger in another.

well adapted there) and from each other by a scattering effect centripetal with regard to the ancestral configuration and linear with regard to each of the separating groups. Such effects are evident in the fossil record as some of the so-called "explosive" stages of evolution in various groups.[15]

A probable example has been given by Bulman (1933), who plotted numbers of species in graptolites and found that periodic great increase in them was regularly followed by a sharp decline. His interpretation is that number of species is more or less directly proportional to adaptive value of the structures developed, and he feels that the subsequent decline is therefore inexplicable. If the opposite point of view is taken, the whole process seems simply explicable. Increase in number of species represents a decline in the adaptive status of the ancestral populations, and consequent centrifugal selection and fragmentation of groups imperfectly adapted but tending more or less toward a variety of different adaptive types under the impulse of linear fractions of this total radiant selection pattern. Some of them achieve a new and perfected adaptive condition and become abundant and successful; the majority do not, and become extinct. The less adaptive phase has many species, and the adaptive phase, few.

Wright (1931) has suggested a figure of speech and a pictorial representation that graphically portray the relationship between selection, structure, and adaptation. The field of possible structural variation is pictured as a landscape with hills and valleys, and the extent and directions of variation in a population can be represented by outlining an area and a shape on the field. Each elevation represents some particular adaptive optimum for the characters and groups under consideration, sharper and higher or broader and lower, according as the adaptation is more or less specific. The direction of positive selection is uphill, of negative selection downhill, and its intensity is proportional to the gradient. The surface may be represented in two dimensions by using contour lines as in topographic maps (Figs. 11-12). The model of centripetal selection is a symmetrical, pointed peak and of centrifugal selection, a complementary negative feature, a basin. Positions on uniform slopes or dip-surfaces have purely linear selec-

---

[15] Some of these records, however, seem only distantly related to this particular phenomenon. The term "explosive" is not altogether happy, since the "explosions" are commonly in process during millions of years.

tion. The whole landscape is a complex of the three elements, none in entirely pure form. To complete the representation of nature, all these elements must be pictured as in almost constant motion—rising, falling, merging, separating, and moving laterally, at times more like a choppy sea than like a static landscape—but the motion is slow and might, after all, be compared with a landscape that is being eroded, rejuvenated, and so forth, rather than with a fluid surface.

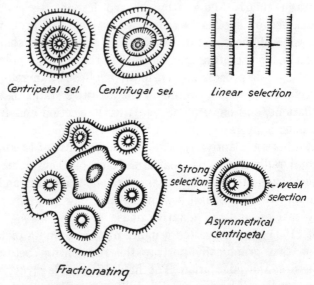

Fig. 11.—Selection landscapes. Contours analogous to those of topographic maps, with hachures placed on downhill side. Direction of selection is uphill, and intensity is proportional to slope.

One aspect of horse evolution, readily explicable in these terms, provides a well-documented, large-scale example that gives more feeling of reality to this rather theoretical discussion (see Fig. 13). An essential element in the progress of the Equidae was the evolution of food habit—there were two main types, browsing and grazing. The most obvious difference is in the height of teeth—browsers having brachyodont teeth, and grazers hypsodont teeth—but there are many other characters directly and indirectly related to these habits. The tooth-crown pattern is more complicated in grazers, the occlusion is different from that of browsers, jaw musculature and movement are somewhat modified, the digestive tract is undoubtedly different, although what the digestive tract of browsing horses was is not known, since they are extinct. Less directly, there are modifications in feet and limbs and throughout the skeleton that are more or less correlated with different habitats—grasslands for the grazers and, in part, woodlands for the browsers.

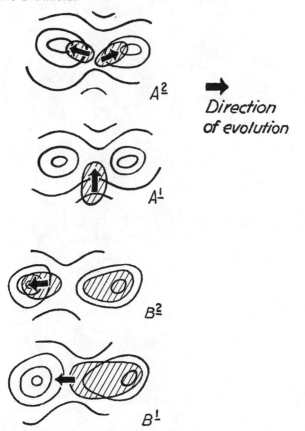

FIG. 12.—Two patterns of phyletic dichotomy; shown on selection contours like those of Fig. 11. Shaded areas represent evolving populations. A, Dichotomy with population advancing and splitting to occupy two different adaptive peaks, both branches progressive; B, dichotomy with marginal, preadaptive variants of ancestral population moving away to occupy adjacent adaptive peak, ancestral group conservative, continuing on same peak, descendant branch progressive.

In the Eocene browsing and grazing represented for the Equidae two well-separated peaks, but only the browsing peak was occupied by members of this family. That peak had moderate centripetal selection, which was asymmetrical, because one kind of variation, on one side, away from the direction of the grazing peak (teeth lower than optimum, and so forth) was more strongly selected against than on the other side, in the direction of the grazing peak.

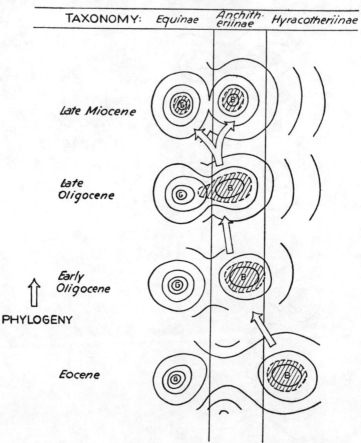

FIG. 13.—Major features of equid phylogeny and taxonomy represented as the movement of populations on a dynamic selection landscape. For fuller explanation see text.

As the animals became larger—throughout the Oligocene, especially —the browsing peak moved toward the grazing peak, because some of the secondary adaptations to large size (such as higher crowns, as previously discussed) were incidentally in the direction of grazing adaptation. Although continuously well adapted in modal type, the population varied farther toward grazing than away from it, because of the asymmetry of the browsing peak. In about the late Oligocene

Determinants of Evolution

and early Miocene the two peaks were close enough and this asymmetrical variation was great enough so that some of the variant animals were on the saddle between the two peaks. These animals were relatively ill-adapted and subject to centrifugal selection in two directions. Those that gained the slope leading to grazing were, with relative suddenness, subjected to strong selection away from browsing. This slope is steeper than those of the browsing peak, and the grazing peak is higher (involves greater and more specific, less easily reversible or branching specialization to a particular mode of life). A segment of the population broke away, structurally, under this selection pressure, climbed the grazing peak with relative rapidity during the Miocene, and by the end of the Miocene occupied its summit. Variants on the browsing slope tended by slight, but in the long run effective, selection pressure to be forced back onto that peak, and the competition on both sides from the two well-adapted groups caused the intermediate, relatively inadaptive animals on the saddle to become extinct. Thereafter browsing and grazing populations were quite distinct, each differentiated in minor ways. The browsing types eventually, at about the end of the Tertiary, failed to become adapted to other shifts in environment and became extinct, while the grazing types persist today.

The structural and phyletic concomitants of this adaptive history are the basis of the usual subfamily arrangement of the Equidae. The small, primitive browsers on the early browsing peak, well isolated from the equid grazing peak, but (as shown by other data, not included in the diagram) still near other peaks occupied by their close allies, such as the early tapirs, titanotheres, and rhinoceroses, are classed as Hyracotheriinae. More advanced browsers, after the equid browsing peak had moved well away from other perissodactyl browsing peaks and nearer the equid grazing peak, are placed in a subfamily, Anchitheriinae. The grazing forms are united in the Equinae.

SUMMARY

Variability.—Pooled or contemporaneous intragroup variability permits rapid differentiation on a low taxonomic level. It cannot suffice for the maintenance of high or moderate rates of evolution over considerable periods of time or for the rise of higher taxonomic categories. Unusually great variability in potent and rapidly evolving stocks is considered rather a result of other evolutionary factors than a cause of rapid evolution. Most evolutionary lines having long-continued moderate rates show relatively constant variability, and sharp increases in the latter are more often related to degeneracy than to advance. In these lines the opposing factors that increase and decrease variability are usually in equilibrium. Maximum rates of evolution are more consistent with low than with high intragroup variability at any one time.

Rate of mutation.—For sustained evolution to occur, mutations are necessary; but it does not follow that rates of evolution are proportional to rates of mutation. Observed moderate and relatively constant mutation rates are consistent with highly variable rates of evolution and would suffice to produce rates of evolution higher than those shown by any continuous paleontological record. There is, however, some inconclusive evidence that occasionally exceptionally high mutation rates have been effective in permitting likewise exceptional evolutionary rates.

Character of mutations.—"Mutation" is a term sometimes applied indiscriminately to different phenomena, and the relationship between mutations and tempo and mode of evolution depends upon the precise sort of mutation involved in the various evolutionary processes. Single mutations with large, fully discrete, localized phenotypic effects are most easily studied; but paleontological and other evidence suggests that these are relatively unimportant at any level of evolution. Saltations, large systemic mutations producing in one step a categorically new type of organism, could theoretically arise by several different genetic processes. Some of these, however, are practically impossible, and others appear to be exceptional and, contrary to repeated claims, to involve only the lower taxonomic levels. Mutations definitely recognizable in known sequences and leading to significant evolutionary progress almost invariably have small or minimum phenotypic effect. Other things being equal, many small mutations are consistent with higher rates of progressive evolution than a few large mutations. These most frequent and most important mutations produce small fluctuating effects in developmental fields. They are particulate and for the most part independent, in themselves, but do not ordinarily show any 1:1 correspondence with single structures or unit characters in a morphologically descriptive sense.

Length of generations.—True evolution takes place between rather than during generations, so that the natural evolutionary rate is per generation rather than per year. In so far as length of generations is an effective factor, temporal rate of evolution should vary inversely with length of generation. No such correlation is evident in the record, and it is concluded that this factor is relatively ineffective. It may, nevertheless, be an influence contributing to exceptionally high rates of evolution, and it seems to have some influence on the rate of small-scale geographic and ecologic differentiation. The relationship between the life cycle and cyclic or secular environmental changes has a less direct but more profound influence on the direction and the mode of evolution.

Size of population.—Size of population is one of the dominant factors in determining tempo and mode of evolution. The chances of fixation of a mutation depend mainly on absolute incidence of mutations, which is determined by mutation rate and population size. In large populations, evolution is extremely slow except under the influence of selection, and then it is almost purely adaptive in type and approximately proportional to selection intensity in rate. In smaller populations selection is less effective, and random effects occur, which become predominant in very small populations. Moderate to large populations tend to be more variable than very small populations, but large populations tend to a genetic equilibrium relatively unfavorable for rapid and sustained progressive evolution. In general, populations of intermediate size are most favorable for this. Maximum rates of evolution may occur in very small populations, but this may be inadaptive and would then lead to extinction—with rare, important exceptions, when it may lead to a radical and rapid adaptive reorientation. It is impossible to make good numerical estimates of the size of extinct populations, but they can often be qualified as probably relatively large or small. Most principles of evolution deduced by paleontologists are based on large populations and thus are not valid generalizations of the whole evolutionary process. Recognition of this limitation removes apparent discrepancies between paleontological and experimental work on evolution and facilitates better interpretation of the fossil record.

Selection.—Three types of theories have denied crucial importance to selection in evolution: Lamarckian, vitalistic, and preadaptation theories. The first fails under the test of experimental results and is generally discarded. The second unnecessarily removes causal study of evolution from the field of science and, whether it is true or not, at least it is not subject to the methods of research or necessary for the isolation of immediate causes. Preadaptation is a real phenomenon, but not a general explanation, and some implications of its generalizations are demonstrably false. Selection is a truly creative force and not solely negative in action. It is one of the crucial determinants of evolution, although under special circumstances it may be ineffective, and the rise of characters indifferent or even opposed to selection is explicable and does not contradict this usually decisive influence.

Selection is a vector having intensity and direction. In large populations very minute selective advantage may be effective in determining the direction of evolution. Selection for a single point optimum would almost eliminate variability in a large population, but other influences do keep variability at a low, but appreciable, relatively constant figure in such circumstances. In adaptive populations strong selection tends to slow or to stop evolutionary change, but it accelerates evolution in inadaptive populations. Extreme reduction of selection has no constant effect on evolutionary rates or directions, since it only relaxes the control of selection and makes other factors decisive.

Direction of selection depends on the relationships of genetic constitution, physical structure, and adaptation in a population. Complex in nature, it has three separable components: centripetal (tending to concentrate a population around a modal type), centrifugal (tending to split a population into divergent variant types), and linear (tending to shift the modal position in one direction). Complete adaptive stability is unusual. Not only changes in physical environment but also slight changes within the population and within individual genetic systems are of almost continuous occurrence, and even under apparently stable conditions the intricate interrelations normally give selection a linear component.

Chapter III: Micro-Evolution, Macro-Evolution, and Mega-Evolution

THE CRUCIAL POINT in population differentiation is that at which discontinuity appears between parts of what was previously a continuous population. The isolating mechanisms that produce or permit discontinuity have been studied at great length and have been shown to be of many different sorts—geographic, ecologic, morphologic, physiologic, and psychologic. The processes are so complex that a complete explanation will not be achieved for a long time, if ever, but the general nature of the solution, or of the possible different solutions, is already sketched. The most important difference of opinion, at present, is between those who believe that discontinuity arises by intensification or combination of the differentiating processes already effective within a potentially or really continuous population and those who maintain that some essentially different factors are involved. This is related to the old but still vital problem of micro-evolution as opposed to macro-evolution.[1] Micro-evolution involves mainly changes within potentially continuous populations, and there is little doubt that its materials are those revealed by genetic experimentation. Macro-evolution involves the rise and divergence of discontinuous groups, and it is still debatable whether it differs in kind or only in degree from micro-evolution. If the two proved to be basically different, the innumerable studies of micro-evolution would become relatively unimportant and would have minor value in the study of evolution as a whole. The great majority of geneticists and zoologists believe that the distinction is only in degree and combination, but the question is kept alive by the energetic dissent of Goldschmidt (1940) and a few others and by certain ways of interpreting the paleontological evidence, for instance by Clark (1930).

The size implications of the prefixes "micro" and "macro" look quite different to an experimental biologist and to a paleontologist. The experimentalists and most neozoologists concentrate on discontinuities of the lowest order, a promising point of attack, but one that

---

[1] The terms are Goldschmidt's, but the idea is old, as he notes. It was fundamental in the work of Linnaeus and others before him.

may produce a myopic outlook. This is a natural division, but it would hardly occur to a paleontologist independently to call it the line between small-scale and large-scale evolution. What a paleontologist would call large-scale evolution, the differentiation between families, orders, classes, and phyla, seems almost unnoted by the experimentalist, or, if he does think of it, he is inclined to dismiss it as inaccessible for profitable study. It is, for instance, surprising to a paleontologist to open an elaborate memoir on the origin of higher categories (Kinsey 1936) and to find that the "higher categories" in question are subgenera, at the highest possible evaluation, and indeed would be called "species" by most paleontologists and many competent neozoologists.

If the term "macro-evolution" is applied to the rise of taxonomic groups that are at or near the minimum level of genetic discontinuity (species and genera), the large-scale evolution studied by the paleontologist might be called "mega-evolution" (a hybrid word, but so is "macro-evolution"). The assumption, as in Goldschmidt's work, that mega-evolution and macro-evolution are the same in all respects is no more justified than the assumption, so violently attacked by Goldschmidt and others, that micro-evolution and macro-evolution differ only in degree. As will be shown, the paleontologist has more reason to believe in a qualitative distinction between macro-evolution and mega-evolution than in one between micro-evolution and macro-evolution.

There are good paleontological examples of the development of discontinuity near the level of macro-evolution. They are not, it is true, as numerous or as well analyzed as might be wished, mainly because paleontologists have been too busy studying higher categories and because so much of their work on lower categories has been so unsound that some of them have questioned whether it is possible for a paleontologist to identify species by any strict and proper definitions (e.g., Scott and Jepsen 1936). Nevertheless, when the materials are available and have been carefully studied, the development of discontinuity between species and genera, and sometimes between still higher categories, so regularly follows one sort of pattern that it is only reasonable to infer that this is normal and that sequences missing from the record would tend to follow much the same pattern. The development of the discontinuity between *Merychippus* and *Hypohippus,* discussed on an earlier page, is one of the most complete examples. There can hardly be a question in such a case that the differentiation of the two lines involved no qualitatively distinct process peculiar to macroevolution. They diverged gradually, almost imperceptibly at first, by the segregation and further modification of genetic factors analogous to those already involved in the store of variability present in their common ancestry.

This and innumerable other examples show beyond reasonable doubt that the horizontal discontinuity between species, genera, and at least the next higher categories can arise by a process that is continuous vertically and that new types on these taxonomic levels often arise gradually at rates and in ways that are comparable to some sorts of subspecific differentiation and have greater results only because they have had longer duration. Two serious questions remain: whether this really is the usual or universal pattern on these levels, and whether it also occurs or is normal for still higher taxonomic categories. The facts are that many species and genera, indeed the majority, do appear suddenly in the record, differing sharply and in many ways from any earlier group, and that this appearance of discontinuity becomes more common the higher the level, until it is virtually universal as regards orders and all higher steps in the taxonomic hierarchy.

The face of the record thus does really suggest normal discontinuity at all levels, most particularly at high levels, and some paleontologists (e.g., Spath and Schindewolf) insist on taking the record at this face value. Others (e.g., Matthew and Osborn) discount this evidence completely and maintain that the breaks neither prove nor suggest that there is any normal mode of evolution other than that seen in continuously evolving and abundantly recorded groups. This essentially paleontological problem is also of crucial interest for all other biologists, and, since there is such a conflict of opinion, nonpaleontologists may choose either to believe the authority who agrees with their prejudices or to discard the evidence as worthless. It is now proposed to re-examine this problem and to seek a reasonable conclusion concordant with the evidence on both sides and with the facts of experimental biology.

MINOR DISCONTINUITIES OF RECORD

On lower taxonomic levels the causes of many apparent discontinuities in the paleontological record are known. An unusually clear, but otherwise typical, example is provided by Brinkmann's data on ammonites; part of them were previously summarized in Chapter I. Some characters in various phyla within the genus *Kosmoceras* show significant linear or nearly linear regressions against thickness of strata: they were progressing definitely and at approximately uniform rates throughout the time recorded. Thus the number of inner ribs

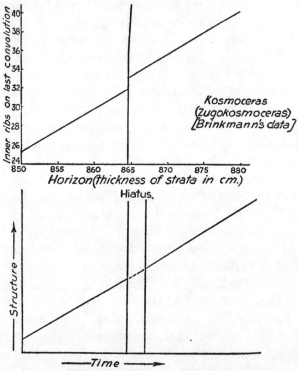

FIG. 14.—An apparent saltation in an ammonite phylum and its interpretation as caused by a depositional hiatus in the beds from which specimens were collected.

on the last convolution in the *Zugokosmoceras* line increased steadily from about 25 to about 40 (as means within the always variable populations) in the 30 centimeters of strata from 850 to 880 of Brinkmann's record (Fig. 14). But at 864.5 the regression line is abruptly offset. The slope remains the same, within the limits of sampling error, but the line jogs upward from about 32 to about 33 (mean values). At the same horizon other characters, although evolving at different rates or in different directions, also are offset and to about the same degree, in proportion to their rates. Also, at this point in the sediments there is a sharp contact, stratigraphically recognizable as a possible hiatus in deposition.

The explanation of the offset in regression is certain on the basis of these facts: there is a span of time not represented by sediments and therefore not by fossils, and the offset represents the amount of evolution of the population during this interval. By separating the regression lines above and below until each lies in the direction of a continuation of the other (Fig. 14), the length of the hiatus can be estimated. It is a period approximately equal to that of the continuous deposition of 2.5 cm. of strata of this deposit.

Such good data are seldom available and almost never so well analyzed, but this phenomenon is very common in the paleontological record. Equally common are cases in which stratigraphic hiatuses are absent or insignificant, but fossils of a given group occur only periodically and hence may appear to represent morphologically discontinuous populations. A large-scale example of such effects is provided by the Puerco and Torrejon Paleocene faunas of New Mexico, similar in facies and occurring in what looks like direct stratigraphic succession. Among the known genera of fossil mammals in these beds, the following have closely related or immediately ancestral and descendant species in the two formations (Simpson 1936b, Matthew 1937, and the collections).

Orders	Puerco	Torrejon
Multituberculata	Kimbetohia	Ptilodus
	Eucosmodon	Eucosmodon
Taeniodonta	Wortmania	Psittacotherium
	Onychodectes	Conoryctes
Carnivora	Loxolophus	Tricentes
	Chriacus	Chriacus
	Protogonodon	Claenodon
	Eoconodon	Triisodon
Condylarthra	Choeroclaenus	Mioclaenus
	Tiznatzinia	Ellipsodon
	Oxyacodon	Protoselene
	Anisonchus	Anisonchus
	Conacodon	Haploconus
	Carsioptychus	Periptychus
	Protogonodon	Tetraclaenodon

Of these fifteen lines the discontinuity of record between the two faunas represents evolution of generic rank in twelve. In the other three, although nominally congeneric in recent revisions, the species of the two faunas are very different and will probably be placed in separate genera when better known. This is remarkably good evidence of the sort used to support the belief that genera and sharply distinct species regularly arise by saltation; but that hypothesis is not really necessary to explain these facts, nor do they support the hypothesis as a real probability.

There are about 170 feet of strata without known fossil mammals between the highest known Puerco and the lowest known Torrejon mammals (Sinclair and Granger 1914) and the difference in level between the more abundant and well-identified specimens is greater. There is no obviously great hiatus in sedimentation, but there are numerous small disconformities and sharp contacts, any one of which may represent a time gap. It was long considered probable (for instance, by Matthew) that these successive faunas were separated by sufficient time to allow for their continuous evolution at rates comparable to those evident in later Tertiary mammals with more continuous records. This inference has recently been brilliantly confirmed by the discovery in Utah of a fauna intermediate between the known Puerco and Torrejon faunas in age (Gazin 1941). Although this fauna (Dragon) is, as yet, considerably smaller than the Puerco and Torrejon faunas, it includes members of eleven of the fifteen lines listed above. Nine of these are morphologically transitional between Puerco and Torrejon types. The other two may be, but are insufficiently known to demonstrate this as a fact. Such discoveries have been so frequent in paleontological history that paleontologists are usually justified in predicting that filling the gaps between sharply distinct successive species and genera of one phylum only requires finding allies of intermediate age.

It is not, however, always true that the evident time gap between successive distinct morphological stages in fossils is sufficient to account for the recorded evolutionary advance without postulating either saltations or unusually high rates of evolution. Some of the exceptions, mostly on high taxonomic levels, seem to be significant and will presently be subjected to special analysis. Others, including most on low taxonomic levels, are probably explicable without these postulates. A good example for which complete proof has subsequently been found is seen in the evolution of the horse as it was interpreted in Europe during the nineteenth century.

One of the first and greatest triumphs of evolutionary paleontology[2] was Kowalevsky's study (1874) of the evolutionary history of the horse and other ungulates. The Old World sequence is [*Hyracotherium*] - *Paleotherium* - *Anchitherium* - [*Hypohippus*] - *Hipparion* - *Equus*. The genera in brackets were not in Kowalevsky's original supposedly direct phylogeny; *Hyracotherium* was believed a collateral rather than a direct ancestor, and *Hypohippus* was not yet distinguished in the old world. Except for a considerable time gap between *Paleotherium* and *Anchitherium* the temporal sequence is almost complete, but there are no transitional forms between the sharply different genera. Thus, the phylogeny seems to have proceeded in a series of abrupt steps, with little evolutionary progress within any one genus and instantaneous jumps of various sizes from one genus to the next. Discoveries in North America during the last sixty years have revealed the true phylogeny and explained the saltations as a peculiarity of the record, not of evolution (Fig. 15). *Hyracotherium* was common to Europe and North America, and in Europe it gave rise to *Paleotherium* and other Eocene genera really well-removed from the ancestry of the horse. *Equus* arose in North America by continuous progressive evolution in which the transitions between genera are not more abrupt than between species within the genera. The four Old World genera subsequent to *Paleotherium* represent as many migrations from America. *Anchitherium*, *Hypohippus*, and *Hipparion* are all from branches off the main line of evolution to *Equus*. Thus the Old World "phylogeny" is doubly indirect, representing discontinuous migration from what were elsewhere continuous sequences and arising, for the most part, from collateral lines, not that leading to *Equus*.[3]

[2] Although almost all the phylogenetic details were later found to be wrong, the achievement was real, and its conclusion fundamentally correct. Kowalevsky deciphered all the essentials of the structural development of the modern horse. He did not identify the precise phyletic lines, because they were not in the materials then available. This is unimportant in comparison with the logical demonstration that the horse had evolved and indication as to how it had evolved.

[3] Yet the inessential part of Kowalevsky's conclusions gained such a hold on European scholars that a few (e.g., Abel) still maintain that *Equus* arose from *Hipparion* and hence by saltation or abnormally rapid evolution—a conclusion almost incredible in the face of the abundant evidence of the *Pliohippus-Equus* transition and universally discarded by the students who have discovered and really familiarized themselves with this evidence.

At best, paleontological data reveal only a very small proportion of the species that have lived on the earth, and the known fossil deposits do not and never can represent anything distantly approaching adequate sampling of the diverse facies of the various zoological realms throughout geological time. Even as regards the most recent and best part of the record, for the Tertiary, and the relatively abundant larger

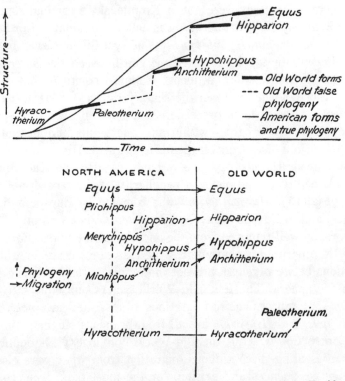

Fig. 15.—Apparent saltatory and true continuous phylogeny of the Equidae. Old World series, a structural sequence of immigrant forms, is saltatory; American series, a true phylogeny, is continuous. American genera not immediately related to the migrating forms are omitted.

land mammals for some of which good sequences are found, the evidence is less than 1 percent of what would be necessary to give continuous knowledge of all the major phyla. In the Paleocene there are fair and rather varied samples from one part of North America, none from Africa or Australia, a single fauna of late Paleocene age and peculiar facies in Asia, and equally poor representation in South America and Europe. For the Eocene there are fair records of limited

Types of Evolution

facies in relatively small portions of Europe and North America, a poorer record of essentially a single facies in what was then as now a marginal part of South America, a few very limited discoveries in Asia, one discovery of latest Eocene age and very aberrant facies in Africa, and none in Australia or many islands (many of which were nevertheless parts of Eocene continents). And so it goes (if not worse) for all types of animals throughout all geological history.

A great deal is known—an amazing amount in view of these limitations—and it would be pointless to emphasize once more the general incompleteness of the paleontological record, except to stress that this incompleteness is an essential datum and that it, as well as the positive data, can be studied with profit. When the record does happen to be good, it commonly shows complete continuity in the rise of such taxonomic categories as species and genera and sometimes, but rarely, in higher groups. When breaks or apparent saltations do occur within lines that are true or structural phyla, frequently they can be shown to be due to one of the two causes now exemplified: to hiatuses in the time record caused by nondeposition of middle strata or fossils and to sampling of migrants instead of main lines. Continued discovery and collecting have the constant tendency to fill in gaps. The known series are steadily becoming more, never less, continuous. It cannot be shown that discontinuity between, let us say, genera has never occurred, but the only rational conclusion from these facts is that no discontinuity is usually found and that there is no paleontological evidence that really tends to prove that there is any. On these levels everything is consistent with the postulate that we are sampling what were once continuous sequences. The collections include about as many continuous series and about as many breaks in various phyla as would be expected from the nature and intensity of the sampling so far accomplished.

MAJOR SYSTEMATIC DISCONTINUITIES OF RECORD

The levels to which these conclusions apply without modification are approximately those discussed as macro-evolution (under that or an equivalent term) by neozoologists and biologists. On still higher levels, those of what is here called "mega-evolution", the inferences might still apply, but caution is enjoined, because here essentially continuous transitional sequences are not merely rare, but are virtually absent.

These large discontinuities are less numerous, so that paleontological examples of their origin should also be less numerous; but their absence is so nearly universal that it cannot, offhand, be imputed entirely to chance and does require some attempt at special explanation, as has been felt by most paleontologists.

Matthew has pointed out (e.g., 1926) that *Hyracotherium* (*Eohippus*) is so nearly a generalized primitive perissodactyl that it could be near the ancestry, if not itself the ancestor, of all the later families of perissodactyls. Knowledge of a nearly continuous sequence leading to the horses and ignorance of smaller or larger parts of sequences leading to other families (tapirs, rhinoceroses, titanotheres, and so forth), at first closely similar, might be due only to chance. But nowhere in the world has any recognizable trace been found of an animal that would close the considerable structural gap between *Hyracotherium* and the most likely ancestral order, the Condylarthra.

This is true of all the thirty-two orders of mammals, and in most cases the break in the record is still more striking than in the case of the perissodactyls, for which a known earlier group does at least provide a good structural ancestry. The earliest and most primitive known members of every order already have the basic ordinal characters, and in no case is an approximately continuous sequence from one order to another known. In most cases the break is so sharp and the gap so large that the origin of the order is speculative and much disputed. Of course the orders all converge backward in time, to different degrees. The earliest known members are much more alike than the latest known members, and there is little doubt, for instance, but that all the highly diverse ungulates did have a common ancestry; but the line making actual connection with such an ancestry is not known in even one instance.

As regards the orders of mammals, Table 15 (pages 108–109) and Fig. 16 give some idea of the inadequacy of the record, both systematic, with regard to the origins of the various groups, and random, within the orders once they have appeared in the fossil record.

Listing of data as to the occurrence of possible ancestry involves subjective judgment as to what constitutes a "possible ancestry," and in some cases opinions differ radically because of the magnitude of the morphological gaps between the bases of ordinal records. These data are also strongly affected by random, and in some cases also systematic, gaps in the records concerning possible ancestors. They do, however, more nearly than any other available information provide objective criteria as to the span within which the orders probably originated.

This regular absence of transitional forms is not confined to mammals, but is an almost universal phenomenon, as has long been noted by paleontologists. It is true of almost all orders of all classes of animals, both vertebrate and invertebrate. A fortiori, it is also true of the classes, themselves, and of the major animal phyla, and it is appar-

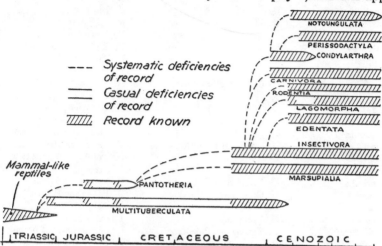

Fig. 16.—Systematic deficiencies of record of mammalian orders. Known spans of some of the orders of mammals with relatively good records and inferred phylogenetic and temporal relations of their systematically lacking origin sequences. No ordinate scale. Abscissal scale arithmetic, approximately proportional to absolute time.

ently also true of analogous categories of plants. Among genera and species some apparent regularity of absence of transitional types is clearly a taxonomic artifact: artificial divisions between taxonomic units are for practical reasons established where random gaps exist. This does not adequately explain the systematic occurrence of the gaps between larger units. In the cases of the gaps that are artifacts, the effect of discovery has been to reveal their random nature and has tended to fill in now one, now another—now from the ancestral, and now from the descendent side. In most cases discoveries relating to the major breaks have produced a more or less tenuous extension backward of the descendent groups, leaving the probable contact with the ancestry a sharp boundary. None of these large breaks has actually

Table
Available Records of Ancestry

Order[a]	Known Occurrence of Possible Ancestry	Recorded Span
Monotremata	U. Triassic	Pleistocene-Recent
Multituberculata*	U. Triassic	U. Triassic-L. Eocene
Triconodonta*	U. Triassic	M.-U. Jurassic
Symmetrodonta*	U. Triassic	U. Jurassic
Pantotheria*	U. Triassic	M.-U. Jurassic
Marsupialia†	U. Jurassic	U. Cretaceous-Recent
Insectivora†	U. Jurassic	U. Cretaceous-Recent
Dermoptera	U. Cretaceous-Paleocene	U. Paleocene-Recent
Chiroptera	U. Cretaceous-Paleocene	[U. Paleocene?] M. Eocene-Recent
Primates	U. Cretaceous-Paleocene	M. Paleocene-Recent
Tillodontia*	U. Cretaceous-Paleocene	U. Paleocene-M. Eocene
Taeniodonta*†	U. Cretaceous	L. Paleocene-U. Eocene
Edentata†	U. Cretaceous-Paleocene	U. Paleocene-Recent
Pholidota	Paleocene	[Oligocene?] Pleistocene-Recent
Lagomorpha†	U. Cretaceous-Paleocene	U. Paleocene-Recent
Rodentia†	U. Cretaceous-Paleocene	U. Paleocene-Recent
Cetacea†	Paleocene	M. Eocene-Recent
Carnivora†	U. Cretaceous	L. Paleocene-Recent
Condylarthra*†	U. Cretaceous-Paleocene	L. Paleocene-U. Eocene
Litopterna*†	Paleocene	U. Paleocene-Pleistocene
Notoungulata*†	Paleocene	U. Paleocene-Pleistocene
Astrapotheria*	Paleocene	L. Eocene-M. Miocene
Tubulidentata	Paleocene-Eocene	[L. Eocene?] L. Pliocene-Recent
Pantodonta*†	Paleocene	M. Paleocene-M. Oligocene
Dinocerata*	Paleocene	U. Paleocene-U. Eocene
Pyrotheria*	Paleocene	L. Eocene-L. Oligocene
Proboscidea†	Paleocene	U. Eocene-Recent
Embrithopoda*	Paleocene	L. Oligocene
Hyracoidea	Paleocene-Eocene	L. Oligocene-Recent
Sirenia†	Paleocene	M. Eocene-Recent
Perissodactyla†	Paleocene	L. Eocene-Recent
Artiodactyla†	Paleocene	L. Eocene-Recent

[a] Extinct orders are marked with an asterisk; orders with long, good records are marked

been filled by real, continuous sequences of fossils, although many of them can be exactly located and the transitions described by inference from the improved record on both sides.

In addition to the fact that they exist, there are other more or less

OF THE MAMMALIAN ORDERS

Continuity of Record (for any considerable part of order)	Remarks
Very short	Probably Australian for much of history, where few fossils are known.
Fair	
Poor	Known from only two ages.
None	All known are of one age.
Poor	Only two ages.
Good & Poor	Good record in America; very defective in Australia.
Good	
Poor	Enormous gap L. Eocene-Recent, probably because of small size and Asiatic occurrence.
Poor	Generally rare fossils, small, fragile, and not likely to occur in normal sediments.
Fair	Uncommon as fossils, but many scattered occurrences.
Good	
Good	
Good	
Poor	Fossil record practically valueless.
Good	Good later than Eocene; practically lacking before.
Good	
Good	
Good	
Good	
Good	
Good	
Fair	
Poor	
Good	
Good	
Fair	
Good	
None	Only one occurrence.
Fair	
Good	
Good	
Good	

with a dagger.

systematic features of these discontinuities of record that call for attention and require explanation.

1. The missing forms (if they existed) were, as a rule, small animals compared with their best-known contemporaries and with their de-

scendants. The earliest members of each mammalian order are smaller than the average for the later members, and a fair inference is that their unknown ancestors were as small or smaller. Among other vertebrates and among the invertebrates there are various irregularities, but for them, also, it is the rule that in the higher taxonomic categories the earliest known representatives are relatively small individuals.

(2.) New groups invariably represent a major ecological adaptive change from their probable ancestors. Although defined morphologically, ordinal and class characteristics of animals, and to less extent those of lower taxonomic units, also represent definite types of physiological functioning and distinctive sorts of environmental co-ordinations. These basic characteristics may later be lost or greatly modified, and the records of such secondary radical ecologic changes are also commonly discontinuous: for example, the return of land animals to the sea and the loss of flight in birds. In a few cases changes analogous to those initiating great taxonomic groups can be followed in lesser groups, an important fact, because it permits inferences as to the probable course of the changes in the unknown sequences. For instance, rodents first appear with definite specialized characters of the dentition, musculature, skull, and so forth, characteristic of the order and never lost in it. Typotheres (extinct South American ungulates) first appear without these rodent specializations (but, of course, with others of their own), and some of them later developed adaptive specializations very similar to those of rodents. The greatest ecological changes are never fully recorded by fossils. Apart from these, all the systematic gaps correspond with important ecological changes, but not all important ecological changes are unrecorded. When there are records, the change is shown to have been gradual, not by saltation.

(3.) The ancestral type, at about the probable time of origin, is often well recorded and representative of probably abundant and widespread groups, while the first representatives of the new type are usually rare as fossils. There are so many exceptions that this is a less secure generalization, but this tendency seems to exist. Devonian crossopterygians are now very well known, but their descendants the amphibians are exceedingly rare fossils until the Pennsylvanian, millions of years after this class originated. Then the amphibians are common, but the first reptiles, in the later Pennsylvanian, are rare. Mammal-like reptiles are common Triassic fossils, but only a few fossil mammals are found before the Paleocene. Primitive primates are not uncommon in the later Paleocene and Eocene, but the earlier representatives of higher types derived from them are among the rarest of fossils. Further significant qualification is necessary, because among the more abundant fossils of generally ancestral groups, the precise line that represented the phyletic ancestry is usually rare or absent, as if these particular ancestors had been rare lines or locally differentiated small populations within abundant groups.

(4.) The major gaps are systematic, but not absolute. Isolated discoveries have frequently been made within what are otherwise unrecorded transitions. These subdivide, but do not fill the gaps. Thus, one of the most perfect of all interclass discoveries, that of *Archaeopteryx* and *Archaeornis,* reveals animals that had not yet perfected the adaptive specializations seen in all Tertiary and Recent birds, but that were birds in basic ecological type: they flew by means of feathers. The most crucial part of the transition was already accomplished. Nevertheless, when such discoveries are made, they are always intermediate in many if not in all respects between the ancestral and the descendent groups. Even in the critical respect of flight structure, *Archaeopteryx* and *Archaeornis* were far more reptilian than any later bird, and aside from this they were about as close to the reptiles as to the birds, or even more reptilian, as Heilmann concluded from his careful study (1926).

(5.) Probably there is always a considerable period of time corresponding with the gap in morphology, taxonomy, and phylogeny. It is impossible to prove that there are no exceptions to this generalization, so that there is some danger that it may represent the statement of an a priori postulate rather than evidence for the postulate; but I believe that it is a valid deduction from the facts. Without prejudice, it can at least be stated that there is sometimes such a period of time and that no facts contradict the possibility. The difficulty arises because slowly evolving and more rapidly evolving lines frequently have the same ancestry, so that later stages in the slow line may be mistaken for the actual ancestry. Thus, the living American opossums have no characters that absolutely preclude the possibility that they were ancestral to all the Australian marsupials, although the real common ancestry was not exactly like the opossum and may have lived 60,000,000 years ago, more or less. All tetrapods were derived ultimately from crossopterygian fishes, and there is a crossopterygian still living (*Latimeria*). The living genus could not, structurally, be the ancestral type, but crossopterygians that could conceivably be the true ancestors were contemporaneous with the earliest amphibians in the late Devonian.

Such facts have been taken to prove the opposite of the present conclusion—that there are no systematic time intervals corresponding with the phylogenetic jumps—but this is an unwarranted inference. The earliest-known Australian marsupial is of the Miocene age and is separated from any known form that is likely (on morphological, paleogeographic, and other grounds) to be a real ancestor by some 30,000,000 years. Crossopterygians more likely to have been tetrapod ancestors occurred in early and middle Devonian times, and the length of the Devonian was on the order of 40,000,000 years. In other and numerous cases there certainly are time intervals. The latest known condylarthra that could be ancestral to perissodactyls occur in the middle Paleocene, and probably they were already too advanced; the earliest known perissodactyls occurred several million years later, in the early Eocene. The latest-known reptiles that were possibly ancestral to birds occur in the Triassic, the earliest-known birds in the middle Jurassic at least 15,000,000 years later, probably more.[4]

6. The appearances of new groups, with unrecorded origin-sequences, frequently coincide with climaxes in the earth's history—crises of mountain building and land emergence. The coincidence with major climaxes, those between the eras, is not as extensive as might be inferred from some texts of historical geology. For instance none of the vertebrate classes can be shown to have originated at the time of a revolution: Agnatha probably in the Ordovician, Placodermi, Chondrichthyes, and Osteichthyes probably in the Silurian, Amphibia in the Devonian, Reptilia probably in the Mississippian—all between and well distant from the Grand Canyon and Appalachian revolutions—Mammalia and

[4] It is commonly assumed by paleontologists, used to the existence of these gaps and inclined to accept them almost too casually, that any group must have originated a period or two earlier than it is actually known. For example, the first mammals (by somewhat arbitrary definition) appear in the Triassic, and it is commonly stated that the class therefore arose in the Permian. But each geological period represents some tens of millions of years (with the sole exception of the brief Quaternary) and this casual predating is as invalid as the opposite inference, that such groups arose in the exact period of their first known remains. The earlier date may occasionally be proved true, but its probability should be judged more carefully by weighing all the data. In the case of the Mammalia, for instance, it is probable that the class arose in the Triassic, and there is no good reason for inferring Permian age.

Aves probably in the Triassic—definitely after the Appalachian revolution and long before the Laramide (classes as in Romer 1933, and names of revolutions those used by Schuchert and Dunbar 1933). It is possible to assume that these and other origins of great taxonomic groups did coincide with revolutions by predating them with respect to the objective record or to relate them to the next previous of the many lesser orogenic periods, but these judgments are purely hypothetical and subjective, not valid deductions from the known evidence. The greatest coincidence of revolutions and important orogenic epochs seems to have been with extinction of old groups, great intercontinental migrations, and expansion, rather than origin, of new groups. There are also nonevolutionary effects: epochs of emergence and orogeny were times of erosion (with respect to what is now land) and so created breaks in the stratigraphic and hence also in the paleontological sequences. Nevertheless, there probably is some real coincidence between tectonic episodes and the rise of new taxonomic groups on the mega-evolutionary level, especially as regards terrestrial animals (or sequences beginning or ending in such), a significant point. For instance, the orders of the four classes of "fishes" and of the class Amphibia show no more probable coincidence of time of origin with times of orogeny than could be due to chance alone, but among reptiles and mammals the proportion of orders that seem to have arisen during times of pronounced emergence and orogeny is greater than can reasonably be ascribed to chance. This is probably true of birds, also, but the fossil record of this class is generally deficient.

Among higher categories the missing major ecological transitions are generally not by dichotomy (both lines surviving), but within a single group already distinct. This generalization has various exceptions and is not entirely secure, but it appears to be another probable tendency. It is quite contrary to the usual diagrams of phylogeny and to the strong convictions of many paleontologists, and therefore it requires brief qualification and defense. Säve-Söderbergh (1934, 1935), concludes that dichotomy is the only phylogenetic pattern in vertebrate evolution, and he makes an entirely new arrangement of Vertebrata along these lines. Even if this premise be granted, the arrangement is impractical; see, for instance, Romer 1936. As diagrams of the affinities of broad groups, such pictures are convenient and, on the whole, valid; but they are subjective simplifications and cannot be taken too literally as guides to the actual genetic facts of phylogeny. For instance, it can be shown that the early Reptilia divided into Mammalia and later Reptilia, and this does not essentially falsify the facts; but it surely is not what happened to real lines of heredity. Reptiles more or less protomammalian in adaptive type became subdivided geographically and ecologically into a large number of local populations. Most of these lines became extinct, but one (or possibly more) continued and achieved the transition to fully mammalian physiology and correlated structure. Each line may have arisen, before this transition, by some sort of dichotomy or by a population division into more than two, but it obscures the essence of the process to say simply that mammals arose from reptiles by dichotomy.

Among lesser taxonomic categories dichotomy in the strictest sense does appear to be a frequent pattern, and many examples are known in essentially continuous paleontological series. It might be more accurate to call the process "polytomy," because it is not confined to splitting into two, and perhaps multiple division is more typical. For instance, on the generic level in the Equidae, *Miohippus* seems to have split at approximately the same time into three generic phyla (*Anchitherium, Archaeohippus,* and *Parahippus*), and *Merychippus* gave rise almost simultaneously to at least six quite distinct and rather complex phyla; four (Stirton 1940) or more (Matthew 1926) of the latter are given separate generic rank. In spite of this and other well-established examples, it is probable that the populations immediately concerned in such minor dichotomies (macro-evolutionary, but not mega-evolutionary) are consistently rarer as fossils than they would be in random sampling of uniformly abundant groups. We find many successive genera lying along single lines, but apparently few that lie at the point of separation of two lines. Thus, Osborn's studies of titanotheres (e.g., 1929) and proboscideans (e.g., 1936, 1942) show many known generic phyla, but virtually no known common ancestor of any two. The number of such common ancestors may be subjectively reduced in Osborn's work, as it was subjectively increased in most earlier work, such as that of Cope. Nevertheless, it is the consensus of paleontologists that such annectent types are unduly rare as known fossils. This is another tendency toward systematic deficiency on a broader, lower scale than those here primarily under discussion. The pattern is different, but the causes are probably related.

EXPLANATIONS OF SYSTEMATIC DISCONTINUITIES OF RECORD

In the early days of evolutionary paleontology it was assumed that the major gaps would be filled in by further discoveries, and even, falsely, that some discoveries had already filled them. As it became more and more evident that the great gaps remained, despite wonderful progress in finding the members of lesser transitional groups and progressive lines, it was no longer satisfactory to impute this absence of objective data entirely to chance. The failure of paleontology to produce such evidence was so keenly felt that a few disillusioned naturalists even decided that the theory of organic evolution, or of general organic continuity of descent, was wrong, after all. Dewar (1931) argued that animals in these major transition periods, if they did really exist, should be the most abundant fossils, rather than the least abundant (as they are), because in their imperfectly adapted state these animals would have been peculiarly liable to fatal accidents.[5]

Disregarding such easily discouraged serious students and ignoring less worthy critics with emotional axes to grind, paleontologists have interpreted the systematic gaps in two ways. One school of thought maintains that the gaps have no meaning for evolution and are entirely a phenomenon of record. Another school maintains that transitional forms never existed. As so often happens, examination will show that neither extreme view is likely to be correct and that the most probable interpretation lies between these two thories, although nearer to the first.

The paleontological evidence for the saltation theory of megaevolution is solely the systematic nature of the breaks in the record. The evidence against this theory is largely indirect, but it is cumulative and in sum is conclusive. According to this theory time intervals should not systematically correspond with the structural breaks; but there is such a correspondence. The breaks should be absolute, and any casual specimens discovered within such time intervals should be definitely on the higher or the lower structural level. In other words, if intermediate types never existed, obviously they would never be found; but

[5] Dewar's whole ingenuous argument is so intricately, appallingly fallacious that one is almost more inclined to treasure it as a perfect thing of its sort than to criticize it. Nevertheless, this and other arguments by its author were taken so seriously by Davies (1937) that he wrote in refutation a whole treatise on paleontological evidence for evolution. We are indebted to Dewar for Davies's fine book, which is both charming and learned and worthy of a less trivial cause.

they are sometimes found. The gaps are not completely filled by such isolated discoveries, so it is always still possible to maintain that saltation did occur earlier or later; but the facts certainly suggest that the sampling of the transition periods has been inadequate, and they lend no real support to the saltation theory. It has previously been pointed out how very unlikely, if not impossible, it is that such major saltations have occurred, according to present understanding of the genetic mechanism. The most nearly concrete suggestion of a mechanism adequate for saltation is that of Goldschmidt (1940), and he quite fails to adduce factual evidence that his postulated mechanism ever has produced or ever really could produce such an effect.

As I see it, then, all the evidence except that of the breaks in the record, is opposed to this theory, which thus merits further serious consideration, in the light of present knowledge, only if no alternative hypothesis is equally or more probable as an explanation of the breaks and is fully consistent with the other data.

The chances of discovering remains of an extinct organism depend chiefly on the structure of the organism (especially its possession of hard parts), its habits, its abundance, the length of time during which it might be buried, the physiographic conditions of its environment, the subsequent history and present exposure of the sediments involved, and the adequacy of search. In the case of many of the invertebrate groups the absence of hard parts makes fossilization highly improbable, and this certainly largely explains the imperfection of the record; but for numerous invertebrates and virtually all the vertebrates, the members of the missing sequences must have been easily preservable as fossils, so that this cannot be seriously advanced as a general explanation. It is true that Rosa (1931, p. 143 and elsewhere) states that the "source species" of phyla, classes, and orders are not fossilizable, but this really gratuitous speculation leads to phylogenetic absurdities; for instance, to Rosa's completely nonsensical conclusion (1931, p. 192) that the latest common ancestor of the marsupials and placentals was not yet structurally a vertebrate.

Habits of animals also have greatly influenced the paleontological record, and to them may be ascribed much of the relative scarcity of fossil birds and higher primates. Similarly, physiographic conditions have deprived us of good records of alpine animals, because they lived

Types of Evolution

in regions of erosion rather than sedimentation. The history of sediments has prevented the finding of many deep-sea fossils, because few of the sediments have been exposed subaerially, of pre-Cambrian fossils in most areas, because the sediments are metamorphosed, and of the fossils in some regions now tropical, because the exposures are weathered and overgrown. Although most of the world has now been summarily examined for fossils and some fields in every continent have been intensively worked, new groups of fossils and even new fossil fields are still being discovered, and unquestionably many still remain to be found.

The major gaps here under consideration have frequently been explained by assuming that the animals in question did live in environments unpropitious for burial, or in regions not yet explored for fossils, or in lands now wholly submerged. None of these explanations, however, can be accepted as general, however important they are as subsidiary factors in separate instances. So many different kinds of mollusks, fishes, reptiles, mammals, and so forth, can hardly all have been subject to limitations of this sort. Such limitations certainly account for many gaps and have had the general effect of uneven attenuation of the record, so that the chances of finding fossils have been reduced; but it is unlikely that they can have caused the extreme reduction of so many sequences that are strongly analogous. It seems probable that these sequences are subject to some other limitations common to all of them and sufficient, together with the subsidiary factors already mentioned, to make the chances of finding them extraordinarily small.

These conditions would be fulfilled if the animals involved in the transitions were relatively few in number and if they were evolving at unusually high rates (Fig. 17). It can be shown that this postulate is consistent with all pertinent facts and, indeed, is almost demanded by them. The effects of these primary or most general factors would be strengthened when they coincided with others also tending, but less systematically, to produce deficiencies in the record: if the transitional animals were small, as they usually were; if they occurred in regions of rigorous physiographic change or of erosion rather than deposition, which must often, but not always, have been true; if they did not continue long in one geographic area or environment, which may have been true of some; if they were not in the same area as their more abundant descendants, which was probably more often true than not; and if their distribution was narrowly localized, which is also a reasonable probability for most of them.

The postulate, usually unexpressed, that rates of evolution have been approximately uniform underlies much paleontological discussion and theory; but it is quite unjustified on the basis of what is recorded

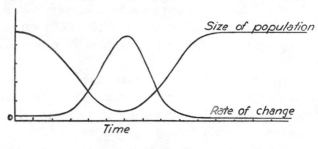

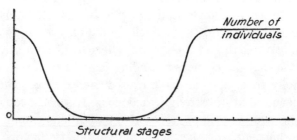

Fig. 17.—Reduction of number of individuals demonstrating major structural changes. In such sequences increased rate of evolution coincides with decreased size of population (upper figure). There are, then, more distinguishable structural stages per unit of time and also fewer individuals per unit of time than in the normal evolution of large populations. The result (lower figure) is an extreme and long-continued reduction in the number of individuals exemplifying each structural stage.

of the facts and theories of population genetics or of what can be inferred as to the gaps in the record. On the whole the horses show about as nearly uniform evolution as any well-recorded group, and they have been cited as evidence for this postulate, but it has been demonstrated in this study that their rate of evolution was not really uniform. Groups, like the Equidae, that have long continuous fossil records and are therefore the paleontologist's main source of opinion as to rates of evolution are exactly those in which the most nearly

Types of Evolution

uniform rates of evolution would be expected. Without important exception they represent large populations living under relatively stable environmental and ecological conditions. In such populations the rate of evolution cannot be more than moderate and will not be likely to fluctuate greatly. In small populations undergoing pronounced shifts in environment and ecology, much higher rates of evolution are possible and much greater fluctuation in rate is probable. From their very nature, such groups do not leave good or continuous fossil records, and it is certainly unwarranted to conclude from this deficiency of record that great fluctuations in rate do not occur.

Attempts to fill the gaps of record by simple extrapolation have frequently been made. In the case of these major gaps extrapolation on the postulate of uniform rates leads to absurdities, although this postulate has been accepted by able students (e.g., Gidley 1923 and elsewhere). The morphological difference between modern opossums and some Cretaceous opossums is slight, but some 60,000,000 years of evolution occurred between them. If the missing pre-Cretaceous sequence changed at a comparable rate, transition from a reptile to an opossum can hardly have taken less than 600,000,000 years; it probably took several times that long—in short it must have occurred in the pre-Cambrian, which is certainly absurd. Or if a structural unit, such as a bat's wing, be studied, it may be found that its recorded rate of evolution is effectively zero. The bat's wing has not essentially progressed since the Middle Eocene, although a few of its nonfunctional elements have degenerated, and it has become more diversified. Extrapolation of this rate in an endeavor to estimate the time of origin from a normal mammalian manus might set that date before the origin of the earth. Attempts to triangulate a date, so to speak, by extrapolation from two lines of common descent frequently also produce contradictions. As a simple example within a known record, extrapolation for hypsodonty from Pliocene equines and anchitheriines would suggest intersection, common ancestry, in about the Upper Eocene (*Epihippus*) stage. The same extrapolation from the later Miocene would suggest about a Middle Miocene (late *Parahippus*) stage. The real intersection is between these, about late Oligocene (*Miohippus*). Use of the most rapidly evolving known line among a number of divergent groups usually produces more consistent results, showing that evolution was usually rapid in the major unrecorded periods, but still often sets incredibly remote dates and suggests that the rate of evolution is still underestimated. For instance, a relatively slow line of rodents, like the Sciurinae, would hardly warrant the assumption that there was intersection with the insectivores later than early Jurassic, and a relatively fast line, like the Hydrochoerinae, might permit placing the date as late as early Cretaceous, but even the latter date seems much too early.

Some idea of probable rates of evolution in the systematic gaps in the record can be gained by comparing the lengths of the gaps to the lengths of the relatively continuous record in the sixteen orders of mammals for which there is fair knowledge. These estimates are highly unreliable, because a large element of personal opinion is involved and no fully objective method of measurement exists as yet. Averages and consistent tendencies should, however, give as good an indication of probabilities as can now be achieved:

TABLE 16

ESTIMATED DURATIONS OF THE ORDERS OF MAMMALS

(In millions of years)

Order	Estimated Length of Unknown Origin-Sequence	Estimated Length of Recorded Sequence	Estimated Total Duration
Marsupialia	55	70	125
Insectivora	55	70	125
Taeniodonta[a]	10	25	35[a]
Edentata	10	50	60
Lagomorpha	35	35	70
Rodentia	20	50	70
Cetacea	20	40	60
Carnivora	10	60	70
Condylarthra[a]	10	25	35[a]
Litopterna[a]	10	50	60[a]
Notoungulata[a]	10	50	60[a]
Pantodonta[a]	10	25	35[a]
Proboscidea	15	35	50
Sirenia	15	40	55
Perissodactyla	10	45	55
Artiodactyla	10	45	55
Means (all)	19	45	64
Means (extinct orders)	10	35	45
Mean percent of total duration	30%	70%	..
Same for extinct orders only	22%	78%	..

[a] Extinct.

Types of Evolution

Even without further study, these figures are more consistent with the systematic recurrence of the great gaps in the record than would be figures based on any close approach to uniform evolutionary rates. Gaps of 10,000,000 years are not unknown between known records of a single sequence, but the vastly greater gaps inferred by many students would be altogether unique if they occurred between known phyletic records.

Regardless of the absolute values of the numbers, these estimates suffice to demonstrate that the basic differentiation of each order took a much shorter time than its later adjustment, spread, and diversification. In the magnitude of structural change involved, this basic part is comparable, on the average, to the subsequent changes, even in the more rapid lines of the orders. The change, for instance, from a carnivore or insectivore to an early cetacean is much more profound than the recorded change from early to late cetaceans; but that from a phenacodont or other condylarth to *Eohippus* is less than from *Eohippus* to a late perissodactyl. It follows that the basic differentiation must have proceeded, on the average, more rapidly than the later recorded evolution, almost surely twice as fast and probably more, quite possibly ten or fifteen times as rapidly in some cases.

According to the theories of population genetics previously summarized, such unusually high rates of evolution are very improbable in large populations and are most consistent with the postulate that the transitional populations were small. This is the hypothesis already shown to be most likely as a primary cause of the systematic deficiency of record, so the two lines of inference reach the same conclusion and strongly re-enforce each other. Other considerations are also consistent. These transitions represent major changes of ecological zone, rather than increasing adaptation to and differentiation within a given zone, as do most of the well-recorded sequences. It is genetically unlikely that such major changes have occurred in large, successful groups that have evolved continuously in one region, but it is likely that they occurred in small unstable groups shifting in location and subject to environmental instability—again conditions strongly reducing the chances of continuous record. Small individual size, although hardly a primary factor, does contribute both to rapidity of evolution and to paucity of fossil discoveries. The paleontological picture of large ancestral populations splitting up into many small groups, which usually become extinct, but sometimes evolve rapidly into radically new types, is also the genetical picture of the situation most likely to produce such new types. These conditions would involve maximum selection, also a strong factor in rapidity of evolution after the more random preadaptive phase that probably initiated each of these major transitions.

The hypothesis that there were unusually high rates of mutation at certain times in the past, which were the times of major evolutionary advance, is attractive to the point of seduction. There is no direct factual evidence for it, however, and it is not a necessary postulate. Moderate mutation rates, more or less like those known in the laboratory, are necessary; but very high rates, even if there were such rates, would not suffice to explain these transitions unless also accompanied by the other special conditions here suggested—and these conditions are adequate without the high rates.

A special form of the mutation-rate hypothesis has been based on the frequent coincidence of rise of new groups and major physical events in the history of the earth: it is supposed that some physical influences especially effective at these times cause a general increase in mutation rates.[6] When there is a coincidence between physical events and evolutionary acceleration beyond any effects probably caused by chance, the groups affected are those whose environment is made unstable or radically altered by the physical events. Such events, acting on such groups, set the stage for preadaptation, for intense selection, for necessary changes in adaptation or extinction, for attenuation and fragmentation of populations—in other words, for just those conditions that do lead to mega-evolution according to the theory here developed. But even without such accelerating physical conditions, moderately progressing large groups can and do reach a stage where transition to another major adaptive type is possible, without necessary connection with physical environmental changes. Some marginal or ecological fragmentation of populations is constantly recurring, and sooner or later one or more such population fragments among many

[6] This idea, already pure speculation, is only farther removed from reality by the attempts that have been made to assign a reason for such an increase—for instance, the action of cosmic rays. There is no clear evidence that cosmic rays were more abundant in particular geologic epochs, or that these were epochs of rapid evolution, or that cosmic rays do increase mutation rates, or that increased mutation rates explain rapid evolution—at the present time this is not even good scientific fantasy, not to speak of scientific inference and theory.

Types of Evolution

make the transition, which is thus more common under major physical environmental stress, but not exclusive to those conditions.

Fenton (1935) has somewhat similarly concluded that revolutions tend more to weaken archaic lines (i.e., to cause extinction of specialized types) than to produce new types. The new types, he says, "remained incipient or rare until the return of stability when, on expanded lands or in widespread seas, they found a chance to develop." Some modification would bring this carefully considered conclusion into agreement with that of the present study. The new types are rare while they are developing, which is more likely but not necessarily true in times of great environmental change, and they spread rapidly after they have achieved a new level of relative stability and their evolution has slowed down and become one of progressive diversification within the now-conquered major ecological zone. Efremov (1935) has also reached much the same sort of explanation of the rarity of transitional forms as fossils, but his explanation of the connection between small populations and rapidity of evolution seems to me almost metaphysical, not explanatory. He supposes that such transitional groups were small because all their energy in the struggle for life was being spent in the rapid adaptive evolution of the organism rather than in its abundant reproduction.

In summary, the theory here developed is that mega-evolution normally occurs among small populations that become preadaptive and evolve continuously (without saltation, but at exceptionally rapid rates) to radically different ecological positions.

The typical pattern involved is probably this: A large population is fragmented into numerous small isolated lines of descent. Within these, inadaptive differentiation and random fixation of mutations occur. Among many such inadaptive lines one or a few are preadaptive, i.e., some of their characters tend to fit them for available ecological stations quite different from those occupied by their immediate ancestors. Such groups are subjected to strong selection pressure and evolve rapidly in the further direction of adaptation to the new status. The very few lines that successfully achieve this perfected adaptation then become abundant and expand widely, at the same time becoming differentiated and specialized on lower levels within the broad new ecological zone.

The most important conditions favoring such a sequence of events are: existence of a widespread group that is evolving progressively, but that has not become genetically immobilized for radical changes in relationship to environment; a varied ecological terrain promoting diversification and fragmentation; shifting instable environmental conditions impeding fixation on any one part of the adaptive landscape or weeding out the groups that do become so fixed; similarly shifting or severe environmental conditions causing extinction of specifically adapted populations and lessening or removing competition within various ecological zones available to the newly arising types.

This is believed to be the typical and, with variations, the general process of mega-evolution, but it is not believed to be confined to that level. Qualitatively similar processes, less only in duration and in the degree of ecological change involved, seem certainly to occur in macro-evolution and even in micro-evolution. The materials for evolution and the factors inducing and directing it are also believed to be the same at all levels and to differ in mega-evolution only in combination and in intensity. From another point of view mega-evolution is, according to this theory, only the sum of a long, continuous series of changes that can be divided taxonomically into horizontal phyletic subdivisions of any size, including subspecies. When a new class, for instance, arises, its transitional stages can be divided into subspecies, species, and so forth, on morphological or genetic criteria. It differs from its ancestry first to a degree that would be ranked as subspecific among contemporaneous groups, then as specific, then generic, and so forth, but the nominal subspecies in this process of successive, cumulative change arise by one particular sort of subspeciation, which is not the most common process of, for instance, geographic subspeciation.

Chapter IV: Low-Rate and High-Rate Lines

THE MOST CASUAL STUDENT of animal history is struck by the fact that while most phyletic lines evolve regularly at rates more or less comparable to those of their allies, here and there appear some lines that seem to have evolved with altogether exceptional rapidity and others that change with such extraordinary slowness that they hardly seem to be evolving at all. Often high-rate, medium-rate, and low-rate lines are quite closely related, and the phenomenon is strikingly illustrated by the common occurrence at one time of what are, in a broad structural way, steps in the successive specialization of a group. Thus the simultaneous occurrence in the recent fauna of a few generalized lemuroids, of many specialized and aberrant lemuroids and many monkeys, and of a few apes and the unique species of man is only explicable on the postulate that the more primitive lemurs evolved more slowly than the average for primates and that the apes and man evolved more rapidly than the average.

Major aspects of the problem of exceptionally rapid rates have been discussed in the preceding chapter and numerous probable examples have been suggested, although in the nature of things the evidence for such rates is largely indirect. The converse problem of exceptionally slow rates is illustrated by more numerous and more direct observations and has long excited special interest and discussion, particularly among paleontologists.

There are many classic examples of low-rate lines: lingulids, Ordovician to Recent; limulids, Triassic to Recent; coelacanths, Devonian to Recent (the lately discovered *Latimeria*); sphenodonts, Triassic to Recent; crocodiles (*sensu stricto*), early Cretaceous to Recent; opossums, late Cretaceous to Recent. The list could be considerably extended. Those mentioned are a few of the so-called living fossils, groups that survive today and that show relatively little change since the very remote time when they first appeared in the fossil record.

Besides other examples of the same sort, there are groups now extinct that were low-rate lines while they lasted, but a survey suggests that the number of these was relatively small—an observation that is important for this problem. It would also be legitimate to add a large number of groups that have become well-differentiated in a comparatively minor way, but that early acquired a fundamental structural type that has been relatively invariable; for instance: turtles, the group as a whole from the Triassic and several recent families and a few genera from the Cretaceous; bats, from the Middle Eocene (at least); armadillos, from Upper Paleocene; rabbits, from Upper Eocene; whales, the order as a whole from the Eocene and modern families mostly from the Miocene. Examples of this sort grade into others like those first given, but they do also introduce a different problem, discussion of which is deferred.

Very low-rate and very high-rate lines may appear as the extremes in an essentially continuous series of evolutionary rates, and all degrees of difference may be found. There is a sort of standard distribution of evolutionary rates, which obviously show considerable variation, and a few rates in any group must lie near the upper limit and a few near the lower limit of the range of rates. If this is all there is to the problem of low and high rates, identification of the factors influencing the greater or less extension of that range will be of great interest, but it will be part of the general study of normal-rate distribution. The apparently exceptional rates will not belong in special categories and will not constitute separate and special evolutionary problems. Investigation of this point, explicitly recognized and so stated, has not previously been undertaken, so far as I know, and it is a necessary preliminary to the study of the causes or concomitants of exceptional rates. Approaching the subject from a different side, these causes and concomitants for high rates have already been discussed to some extent. It remains to relate this discussion to some broader aspects of rate distribution and to discuss low rates in more detail, for these do involve exceptional and special factors beyond those of the standard rate range, as will be shown to be true.

DISTRIBUTIONS OF RATES OF EVOLUTION

In Chapter I of this study these facts were exemplified: (*a*) related lines of descent commonly differ in evolutionary rate; (*b*) within larger groups, such as classes, there is an average or modal rate of evolution that is typical of the group; and (*c*) these average rates differ greatly from one group to another. Here is seen again the important distinction between intragroup and intergroup variation, the variate in this case being the rate of evolution, individual values of the rates within single phyletic lines—and the groups being large taxonomic units comprising many phyletic lines of ultimate common origin. The average or group rate for the Mammalia is much greater than that for the Pelecypoda, but a low rate for a single line among mammals may be lower than many pelecypod rates.

Discussion of low and high evolutionary rates requires designation of the standard of reference. Individual rates may be low or high relative to the average for the group to which they belong, and group averages may be low or high relative to the average of another group or to the average for a number of groups taken together. The ancestry of *Didelphis* has a low rate relative to marsupials or mammals as a whole. Mammals have a high rate relative to the animal kingdom as a whole. That the *Didelphis* phylum has a moderate to high rate relative to the Pelecypoda is a meaningless statement unless viewed as a corollary of the fact that mammals have a high group rate relative to pelecypods.

The similarity of different survivorship curves, as previously demonstrated for such disparate groups as pelecypods and mammalian carnivores (Chapter I), suggests that there is a standard pattern of survivorship that is essentially the same for many, probably for all, taxonomic units, the various units differing markedly from each other in mean absolute values, but little in the pattern of distribution within units. Since survivorship is believed to be sufficiently correlated (negatively) with rate of evolution to serve as an approximate, indirect measure of the latter, this suggests that there is also a standard distribution pattern of rates of evolution within groups and that they tend to have the same sort of intragroup relative dispersion of rates in spite of the great intergroup differences in absolute values. If the negative correlation of survivorship and rate of evolution were perfect —which, of course, it is not, but the hypothesis serves to set up a model illustrative of tendencies—then the intra-group distributions of rates of evolution would follow patterns like those shown in Fig. 18.

These distributions are strongly peaked (leptokurtic with respect to the normal curve) and are also strongly asymmetrical (negatively skewed with respect to the normal curve). How far these striking features of the model or postulate are present in real (practically unobtainable) distributions of organism rates of evolution cannot be positively asserted, but close correspondence would be expected and may reasonably be assumed in the absence of a more direct approach.

The peakedness of these distributions means that among many allied phyla most of them tend to evolve at rather similar rates. In Pelecypoda and Carnivora from one-third to one-half of the genera are

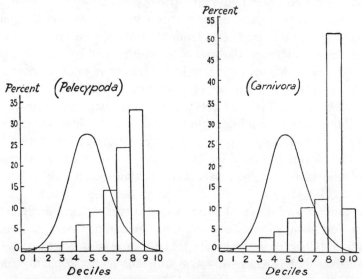

Fig. 18.—Frequency distributions of rates of evolution in genera of pelecypods and land carnivores. Histograms based on survivorship of extinct genera (see Figs. 5-6) and on the postulate of —1.00 correlation of survivorship and rates of evolution. Ranges (abscissal scale) divided into deciles to produce histograms strictly comparable in form although absolute rates are very different. Normal curves equal in area to the histograms drawn for comparison.

evolving at rates included in only one-tenth of the whole range of variation in rates, a concentration greater than is probable in chance sampling of a normal distribution, in which the middle tenth of the range includes only about one-fourth of the total frequency. This tendency is evident in most paleontological series, and empirical generalization from observation agrees closely with the results of this more theoretical approach.

The asymmetry of the theoretical distribution means that within a group of allied phyla rates higher than the mode are less common than those below the mode; that more phyla evolve at rates near the maximum than at rates near the minimum. In both the models used as examples the mode is in the ninth of ten equal-range classes (abscissal deciles) and 65-75 percent of the frequencies are above the two middle classes (fifth and sixth) but only 9-10 percent are above the modal class. In the normal distribution only 26 percent of the total frequency is above the two middle decile classes, and half the frequency is above the mode. This asymmetry, also, is a reasonable generalization from direct observation of the fossil record, although it has not been so clearly noticed. Both tendencies were suggested by Matthew (1914) in less explicit terms and as subjective opinions, but opinions backed by almost unparalleled knowledge of the objective data of vertebrate paleontology.

These models and the standard pattern of rate distribution inferred from them are based on survivorship data for extinct genera only, for those that had run their whole evolutionary course and therefore have determinate survivorship. Similar data can be obtained by plotting percentages of known genera that have arisen in the various geological epochs and survive today. If rates of evolution of living genera have been distributed as have those of extinct genera, or, in statistical terms, if the rate distributions sample the same populations, the resulting curves will nearly coincide with those based on extinct genera. It has been shown (Chapter I) that curves derived in the two different ways have, in fact, pronounced discrepancies.

These discrepancies show up still more clearly, for present purposes, when a comparison is made between the observed living fauna and the fauna that would be expected on the basis of survivorship among extinct genera. Data of this sort for pelecypods are given in the following table. The survivorship curves for extinct genera permit calculation of the percentage of genera arising in any given epoch or time interval that would be expected to survive for the length of time represented by the span from their appearance to the present time, column A, "Expected," in Table 17. Realization of such survivorship is reckoned from the percentage of known fossil genera that first appeared in each stated time interval and survive today given in Column B, "Realized," in Table 17. The difference between the expected and the realized percentages is the discrepancy to be accounted for.

The accompanying graphs (Fig. 19) represent a different way of considering and representing the same phenomena. They show, not

TABLE 17

EXPECTED AND REALIZED SURVIVAL INTO THE RECENT OF BROAD PELECYPOD GENERA OF VARIOUS AGES

Time of Appearance (In millions of years previous to present)	A. Percentage of Expected Survival, from Survivorship in Extinct Genera	B. Percentage of Realized Survival	Difference B-A
25	92	78	−14
50	63	62	− 1
100	24	39	15
150	8	30	22
200	2	24	22
250	1	19	18
300	0	13	13
350	0	8	8
400	0	4	4
450	0	0	0

the percentage of survival expected and realized for given time intervals, but the expected and realized age composition of the Recent fauna as a result of such survival. For this purpose the Recent fauna includes only genera that do occur as fossils. The data are the percentages of these recent genera that first appeared at given times, "Realized" (solid lines in the graphs) and the corresponding percentages as they would be if survival had followed the curves derived from survivorship in extinct genera, "Expected" (broken lines in the graphs).

The Recent carnivore fauna is less rich than would be expected on the basis of the standard survivorship curve, but as Fig. 19 shows, its age composition is very near expectation. The average age of living genera is slightly greater than would be expected, but the discrepancy is so small that it may be due wholly to chance and to inaccuracy of record. No living genus has survived longer than the maximum known for extinct genera. All that has happened is that the rate of extinction, evolution, or both during late Pliocene to Recent was greater than it had been before that, but its effect was not markedly differential and was about the same as an average, on genera of all ages.

The Recent pelecypod fauna, on the other hand, is considerably richer than would be expected, and its age composition is markedly different from expectation. Fewer young genera (Tertiary) survive than would be expected. Extinction and evolution have affected them more than old genera. On the other hand, many more old genera survive than would be expected, a discrepancy greatest for Triassic genera, but evident for all pre-Cretaceous periods. Moreover, a number of living genera have already survived much longer than did any genus now extinct. Defining genera very broadly, no pelecypod genus now

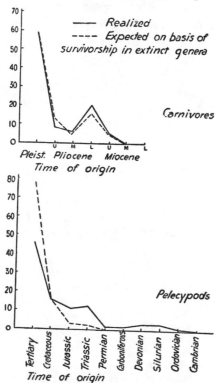

Fig. 19.—Realized and expected age compositions of recent pelecypod and land carnivore faunas. Percentage of living genera arising (first appearing in fossil record) in stated epochs and periods as observed in the recent fauna (counting only genera that are known as fossils) and as would have been predicted on the basis of survivorship in extinct genera.

extinct appears to have lived for more than about 275,000,000 years, which corresponds with the lower limit of the standard distribution of evolutionary rates in this class. But some living genera, defined with equal broadness, have endured for upwards of 400,000,000 years without becoming extinct.

Such genera clearly belong quite outside the standard rate pattern. They are not merely the most slowly evolving individual lines of the standard rate distribution. They reveal a distinctly different set, or statistical population, of rates.

All the pelecypod phyla (about ten) that have survived with no more than broadest generic advance since about the Carboniferous or earlier belong, *ipso facto*, to this special, non-standard rate distribution. Examples familiar as living bivalves are the collective genera *Nucula*, *Leda*, *Modiola*, *Avicula* (or *Pteria*), *Lima*, and *Ostrea*. The

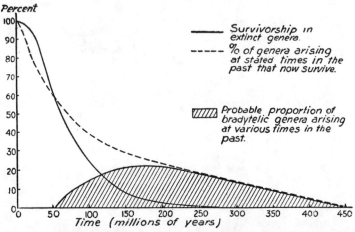

Fig. 20.—Bradytely in pelecypods. Survivorship in extinct genera, continuous, unshaded curve, and analogous curve based on recent genera, broken curve, see Fig. 5. The shaded area represents the excess of realized over expected survivorship caused by the nonextinction of bradytelic phyla and so is approximately proportional to the rise of bradytelic lines at the stated times in the past. (The apparent absence of bradytelic lines arising in the last 50,000,000 years may be real, but is probably an artifact caused by differential extinction of horotelic genera in this period, the extent of which cannot be estimated from the available data.)

great excess of one curve (broken line, Fig. 20) over the other (solid line) for lesser spans and later times demonstrates that these nonstandard low-rate lines continued to arise long after the Carboniferous and even to originate more frequently, at least through the Jurassic. Here, however, it is impossible to say of any specified genus that survives today whether it is a relatively slow line of the standard distribution, destined to disappear or to evolve into something else within a few million years, or belongs in the nonstandard group and is not destined to change for an unpredictably long time. By definition (but the definition clearly represents a real distinction) known extinct genera, whenever they arose, belong to the standard rate distribution.

The conclusion is inescapable that the present pelecypod fauna is the resultant of at least two decidedly different rate distributions. Since there are qualitatively different rate distributions, designations for them are necessary in order to maintain the distinction and to discuss them without ambiguity. It is proposed to refer to the standard rate distribution and the phyla belonging to it as "horotelic" and to refer to the nonstandard, low-rate distribution and its phyla as "bradytelic."[1] A distinction is to be made between slow horotelic rates and bradytelic rates, also slow but essentially a different phenomenon, as demonstrated by the marked difference in survivorship pattern from which the existence of bradytelic rates has been deduced.

Carnivores as a whole seem to have been horotelic, but pelecypods do not agree with prediction on the hypothesis that all are horotelic, and the Recent fauna is characterized by the survival of numerous bradytelic lines.

Indeed, among pelecypods I find no evidence that a bradytelic line has ever become extinct; they seem to be virtually immortal. It is probable that some bradytelic phyla have become extinct in other classes, but this is exceptional. No instance has been found of a bradytelic line that has become horotelic as a whole, but it would be very difficult to verify this phenomenon if it should occur. On the other hand, there are numerous examples of bradytelic groups that have given rise (by branching) to manifestly horotelic and even fast horotelic groups, e.g., the rapidly evolving *Gryphea*, which arose from the bradytelic *Ostrea*.

Another peculiarity that can be inferred from the pelecypod data is that the number of bradytelic genera arising within a given span seems to have increased as time went on. The increase is not marked, and it is not quite certain that there was one, but it is probable.

The study of survivorship cannot, by the nature of the method, reveal whether there are also phyla that belong to a nonstandard rate

---

[1] This subject and that of evolution in general are already so cluttered with unnecessary and personal technical terms that addition to their number is inexcusable without the strongest sort of impulsion. I rewrote this chapter twice in the effort to make it clear without departing from the scientific vernacular, but each time readers found it ambiguous and requested the use of special terms, so that it was finally rewritten again with these two terms included.

distribution, in which rates average much higher than for allied horotelic lines, analogous to the much lower average for the nonstandard bradytelic distribution. Nevertheless, there is good reason to believe that there are groups that have evolved at rates distinctly above those well recorded for horotelic lines and qualitatively different in being affected by a characteristically different combination of evolutionary factors. Such lines may be called "tachytelic" and their nature was a principal subject of the preceding chapter. One aspect of the nature of tachytelic phyla is that they cannot long endure as such, but must soon either become extinct or become horotelic or bradytelic. In subsequent discussion of bradytelic evolution it will be shown that bradytelic lines probably normally arise from tachytelic.

As also discussed in the last chapter, another characteristic of tachytelic lines is that they must be rare or lacking as fossils. Thus, their identification is largely a matter of inference from known ancestors and descendants, and objective examples of phyla demonstrably in the tachytelic phase are necessarily very few. Under unusual conditions they can, nevertheless, be preserved, and some examples are known. One of the clearest of these is the evolution of *Valenciennesia* as interpreted by Gorjanovic-Kramberger (1901 and 1923). This phylum was an offshoot of *Limnaea,* a common, spired, pulmonate snail living in clear fresh water, either a slow horotelic or bradytelic genus (adequate analysis of the gastropods in this respect has not yet been made). *Valenciennesia* was a bowl-shaped, limpet-like form, the shell very thin, but strongly ribbed, with a pulmonary groove (absent in *Limnaea*), that lived in mud in brackish water. By the usual criteria, the difference between *Limnaea* and *Valenciennesia* is that of family rank (indeed, they were generally placed in different suborders before this transition was demonstrated). This extraordinary transition occurred during a part of the early Pliocene, a span during which a single species of horses (*Hipparion gracile*) continued without appreciable change on the adjacent lands. Yet horses, like mammals in general, had much higher average evolutionary rates than gastropods, or their pulmonate order, or the Limnaeidae, nearest relatives of *Valenciennesia*. The extremely high rate of the line leading to this genus lies quite outside the horotelic distribution for any larger taxonomic grouping including it, and this phylum is evidently tachytelic.

The special conditions that led to this unusually adequate record of a tachytelic line are identifiable. The animals lived in land-locked basins from which they could not migrate, so that their whole evolution occurred in a limited region that is now dry land and accessible for collecting. Their mud-living habits made them particularly liable to burial and fossilization, which was the usual fate of a dead individual rather than an exceptional accident as it was with most animals. There was virtually continuous deposition and no erosion while their evolution was occurring. Subsequent erosion has sufficed to expose the strata, but not to destroy any great part of them.[2]

FACTORS OF BRADYTELY

A serious difficulty in the study of bradytely as a phenomenon distinct from slow horotelic evolution is the impossibility in many cases of demonstrating that some one particular line is or is not bradytelic. Thus, among pelecypods generic groups or collective genera surviving from the Carboniferous can be explicitly called bradytelic with sufficient confidence, but not those surviving from the Triassic. Of the twenty living genera that have close allies and prototypes appearing in the Triassic, the survivorship data establish it as probable that two are horotelic and possible that none is or that as many as six are. Hence eighteen probably are bradytelic, but all may be or as few as fourteen may be.[3] It can be said that most Triassic-Recent generic groups are bradytelic, and study may be based on the probability that factors affecting most of them are likely to have a bearing on bradytely, but it is impossible to say of any one genus whether it is bradytelic or horotelic. In other groups, such as late Cretaceous-Recent pelecypods, it is impossible to say not only what genera are bradytelic but also whether any are, so that no study of the problem can be based on these (see Fig. 20).

[2] It is possible that even this unusual example of recorded tachytelic evolution is due to incorrect interpretation, as Basse (1938) maintains. Basse's objection is based on a supposed *Valenciennesia* from marine Cretaceous beds in Madagascar, but he had only one specimen, inadequately preserved or prepared to show more than a superficial resemblance to true *Valenciennesia*. Some doubt may be cast on Gorjanovic-Kramberger's conclusions, but this evidence is quite insufficient to disprove them. The example of tachytelic evolution still seems probably valid, and should it prove invalid, the occurrence of this sort of evolution would, nevertheless, be unchallenged.

[3] As at numerous other points in this study, these statements are based on calculations of statistical probability and are not casual judgments. It seems unnecessary and would make the publication unduly long and complex to include the many calculations and full tables of data on which conclusions are based.

Recognition of the existence of bradytely in a given group and identification of particular bradytelic genera can only follow previous determination of the horotelic rate distribution for the same group. Previous studies of slow evolution, such as the classic work by Ruedemann (1918, 1922a, and 1922b) on what he calls "arrested evolution," confused what now appear to be three distinguishable phenomena; their causes or concomitants cannot be assumed to be the same and seem to me to differ, at least in degree. First, there are the groups with the modes of their horotelic rate distributions low in comparison with the animal kingdom as a whole; second, phyla with low rates within the horotelic distribution for their group; and third, bradytelic phyla. Use of an arbitrary a priori criterion of "arrested evolution" not only confuses these but also is likely, in certain groups, to include some phyla that were really evolving at average or even at fast rates for their own group rate distribution. Thus Ruedemann defined as "persistent types," examples of "arrested evolution," all genera that pass through more than two geologic periods. But among pelecypods this does not necessarily require less than the average rate of evolution, and, on the other hand, it excludes all but a few fishes, practically all reptiles, and all mammals, despite the great probability that bradytely occurs in all these classes. Ruedemann's special class of "immortal types," those broad genera that range from Paleozoic to Recent, probably is composed entirely of bradytelic phyla, with a few possible exceptions in invertebrate groups with the lowest horotelic modes, but it certainly excludes many and probably the great majority of bradytelic lines.

Because the necessary estimation of horotelic distributions is a long and difficult task that has hardly been begun, it would now be premature to offer a detailed list of factors involved in bradytely. What can be done is not much more than to set up a series of hypotheses as to factors that may be involved, on theoretical grounds or from preliminary observation, and to offer these as lines of attack and for future proof or rejection.

Of the factors supposed by Ruedemann and others to bear on persistence or slow evolution in any of the three different senses noted above, a few appear to have probable relationship to bradytely. Some seem to be related rather vaguely to slow evolution of other sorts or in general, some must be considered unimportant aspects or symptoms of more inclusive factors and principles, and some seem to me to be doubtful, erroneous, or only associated with slow evolution by coincidence. Analysis of the visible characters of low-rate and bradytelic phyla is essential and has not yet been accomplished in any detailed and reliable way, despite the abundant literature of the subject; but ultimately these characters must be considered superficial—not the goal of such study, but only data for inference as to underlying genetic or other fundamental factors within the organism and in the organism-environment complex.

Among the individual characters supposed to be associated with slow evolution are adaptation to stable environments, sessile growth, small size, nocturnal or secretive habits, great individual vitality, high fecundity, carrion-feeding, asexual reproduction, occupation of marginal ecological niches, and long individual life span (mainly from Ruedemann, paraphrased, rearranged, and supplemented). All these characters occur in some slowly evolving groups, but all also occur in groups with normal or rapid rates of evolution, and in the terms here proposed I see no present evidence that any one, or a combination of them, is more likely to accompany bradytelic than horotelic evolution.

Looking at the matter in a different way, from the point of view of influences that might theoretically be conducive to slow evolution, the principal individual characters that come to mind as possibilities are low mutation rate, long life span, and asexual reproduction (as inhibiting genetic variability). At present it seems unlikely that any of these is important. Low mutation rate has been the most popular genetic explanation (e.g., Dobzhansky 1941, p. 41), but there does not appear to be any good experimental support for this as yet. Some indirect evidence opposes it. Bradytelic lines probably normally arise from tachytelic and often give rise to horotelic or tachytelic offshoots that live at the same time and that have the same initial genetic structure as the bradytelic group. In the few living probably bradytelic groups for which there are concrete data on phenotypic variation, this appears to be fully comparable in extent and nature with the variation of allied horotelic phyla. Although a very high mutation rate in a very small population would certainly prevent bradytely, it seems doubtful whether an exceptionally low mutation rate is necessary for bradytely or sufficient, in itself, to produce bradytely. As for longevity and asexual reproduction, the majority of known or suspected bradytelic groups do not have longer life cycles than their horotelic allies, and most of them reproduce sexually. Another possible and suggested individual factor, fecundity, would probably tend to promote rapid, not slow, evolution if accompanied by high mortality and would produce large populations, a more important factor, if mortality were not high.

Although, again, the factual data are wholly inadequate, population size must almost necessarily be a primary factor in bradytely. Bradytely involves not only exceptionally low rates of evolution but also survival for extraordinarily long periods of time. Such survival with little change implies the almost complete elimination of random modifications and, of course, excludes chance extinction, requirements most unlikely to be met continuously for such long periods unless the population is large. For many groups of particularly slow evolution the existence of large breeding populations has been demonstrated or is a probable inference. There are apparent and possible exceptions, notably *Latimeria*, of which only one individual has ever been found, and *Sphenodon*, the population of which is now near the vanishing point. But we have no reason to believe that *Latimeria* may not have been or may not even now be abundant in its own habitat, which is little known. As regards *Sphenodon*, its reduction in numbers is recent, and it was abundant so few generations ago that time has not yet sufficed for any pronounced small-population effects.

It is not necessary or helpful to postulate widespread geographic occurrence, but only a large interbreeding group at any one place. If it could be demonstrated that the population of a bradytelic phylum had been small for a long period, then it would be necessary to assume that its mutation rate was extremely low; but it seems probable that no such case has occurred and certain that it is not typical of bradytely.

Bradytelic groups seem to have some features of phyletic history or evolutionary pattern in common. In the first place, they are not primitive, as might be supposed, when they become bradytelic; on the contrary they are then usually progressive and relatively high types. It is only their long persistence after their allies have gone on to explore other avenues of structural modification that gives them an archaic aspect and makes some of their characters primitive in comparison with their contemporaries in late phases of their history. The lingulids had a fairly high type of invertebrate structures for the Ordovician, and the limulids were advanced for the Triassic. The sphenodonts were nearly as progressive as any Triassic reptiles, and so were the crocodiles in the Cretaceous. The opossums were about as advanced as any mammals in the Cretaceous. All these groups are primitive in the recent fauna, because they have been passed in the race, not because they had an initial handicap. This observation becomes still more striking if some of the less clear-cut cases of exceptionally slow (in these instances not surely bradytelic) evolution are included. For instance, the very slowly evolving armadillos were already extraordinarily specialized at the beginning of the Tertiary, and the bats, among the most specialized creatures in the entire animal kingdom, were as advanced in the Eocene as they are now, aside from unimportant details.

Ruedemann (1918), Rosa (1931), and others have already pointed out this peculiarity in other words. Ruedemann says that although "persistent types" are now "weak forms, they were . . . originally the most vigorous stocks." Racial vigor and weakness seem to me vague metaphorical terms used for a variety of essentially different evolutionary phenomena, and I question whether most bradytelic groups are now "weak" in an acceptable sense of the word, but Ruedemann means essentially that what I call "bradytelic" lines arise by relatively rapid adaptive change, and this is probably true. Rosa's generalizations seem to me almost completely untenable, but he does point out that the lines that later evolved most slowly must have evolved rapidly at some previous time, not only more rapidly than in their own subsequent history but also more rapidly than most of their early contemporaries. He calls these "precocious branches," but the term is confusing in its application to groups that are archaic and have been slowly evolving during most of their known history.

The conclusion that most of the known exceptionally low-rate lines must at some previous time have been high-rate lines seems inescapable. It is this phenomenon that makes most palpably absurd the results of extrapolation on the hypothesis of uniform rates of evolution. This conclusion receives further support from examples on a smaller scale, or within more restricted limits, of paleontological records that show rapid evolution followed by a longer period of relatively slow evolution. Numerous examples of so-called "explosive" evolution show such a period of rapid divergence followed by a sharp drop in evolutionary rate. To specify only one instance among those familiar to me, the notoungulates were evolving very rapidly in the Eocene and the Oligocene. In notoungulate lines paralleling the horses, for example, their (approximately) early Eocene to early Oligocene progress is comparable to that of the horses from early Eocene to early Pliocene. But the relatively few notoungulate phyla that survived into the later Tertiary made no further advance of noteworthy importance and evolved at moderate to low rates.

It is probably in the organism-environment relationship that the principal factors in bradytely and the reason for this sort of sudden drop in rates are to be sought. Most low-rate lines and, specifically, most or all bradytelic phyla seem to be highly and particularly adapted to some ecological position or zone with broad but rather rigid selective limits. They have few distinctive peculiarities that are not manifestly or probably adaptive. This contradicts a superficial impression that because an opossum, for instance, is primitive (i.e., generally archaic in structure in comparison with most of the recent Mammalia) it is also generalized and not specifically adaptive. On the contrary, everyone who is familiar with opossums in the field knows that they are among the most perfectly and successfully adapted of all mammalian types. Their ecological field is broad, and the adaptation is not of the extreme "peaked" sort, of, for instance, animals dependent on one sort of food, such as the panda or koala. For this reason the adaptation does not necessarily involve loss of plasticity, and the low-rate line remains capable of throwing off branches that invade narrower or more fluctuating ecological niches and that evolve more rapidly in this new adjustment. The opossums have repeatedly done this, as have many bradytelic groups. The bradytelic group remains and survives as an unprogressive nucleus of larger population, and the more rapidly evolving lines, more at the mercy of changing conditions, often become extinct.

This relationship involves an equilibrium between tolerance and cyclic environmental change. At one end of the sequence are animals that tolerate almost no alteration of external conditions, but live in a virtually unvarying environmental niche, and at the other end are animals that tolerate extreme cyclic variations in temperature, salinity, and so forth, and live in environments characterized by such cycles. These periodically fluctuating environments must, however, be no less permanent, even though more variable, if bradytelic lines are to be found in them. Thus, an animal that became adapted to continuous cold during the Pleistocene would either become extinct or rapidly evolve a different physiological adjustment in post-Pleistocene times. But an animal that developed wide temperature tolerance, which is no less an adaptation than is adjustment to a particular part of the temperature range, would tend to continue unchanged.

Bradytelic groups do not occur in impermanent physiographic surroundings such as lakes or, to less degree, deserts or in extreme climatic zones such as the far Arctic. They are usually to be found in the sea, its strand, the shifting but essentially permanent major rivers, or the more slowly shifting and almost equally permanent great forest belts. They are especially common in the most nearly permanent of climatic zones, the subtropical, whence they ebb and flow into the tropical or into the temperate zone as occasion arises.[4]

The final and probably the most fundamental factor in the relationship is that bradytelic groups are so well adapted to a particular, continuously available environment that almost any mutation occurring in them must be disadvantageous. If a mutation is advantageous in another available environment, then it may become fixed in part of the population and a branch phylum may arise, but the central group will remain. In this group almost any change will be opposed by some selection pressure. In large populations even very small selection pressures are the dominant factors in evolution, whether for conserving the adaptive level already attained, as under these postulated conditions, or for inducing change. It is theoretically probable and it is consistent with the observed facts that bradytely results from the equilibrium of large breeding populations of animals specifically adapted to a continuously available environment that is relatively invariable or has a rhythmic variation that corresponds with adaptive tolerance in the population. The exact coincidence of such factors is evidently unusual among animals, because bradytelic phyla are fewer

[4] It may be considered anomalous that most deep-sea fishes seem to be of relatively recent origin, despite the fact that this would appear to be the most invariable of all environments. The anomaly is, however, explicable. The physical invariability of this environment is conducive to loss of tolerance so that any slight changes in it, such as have undoubtedly occurred, would cause widespread extinction of its too well-adapted inhabitants. Moreover, their relatively recent origin does not mean that the deep-sea fishes are not bradytelic; many of them may be, but it would take a few million years to prove that they are. It is also likely that some, not bradytelic, suffer from small-population effects.

than others. Approximation of such conditions without attainment of full equilibrium is probably more usual and would produce slow horotelic, rather than distinctively bradytelic, evolution. Although the rate distributions are qualitatively distinct, it is unlikely that they are discontinuous, and one undoubtedly grades into the other.

THE SURVIVAL OF THE UNSPECIALIZED

It is not universal, but it is very common, for the genera in a given taxonomic unit, such as a family or a superfamily, to represent comparatively slight, apparently inadaptive and inconsequential variations of a limited number of structural grades. These grades usually follow in serial order, corresponding to adaptive zones that are progressively more narrowly defined, more specific in their demands on the organism,

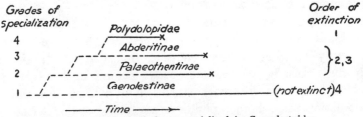

FIG. 21.—Survival of the unspecialized in Caenolestoidea.

and more subject to secular change. The lowest grade, that is, the broadest and most stable adaptive zone, is frequently occupied by a bradytelic, sometimes by a slow horotelic, group whose early members are the actual and later members the structural ancestors of the more divergent lines. The transitions from one grade to the next are rarely recorded by fossils, perhaps never fully recorded, and are normally tachytelic. Within the higher grades and sometimes also the lowest, evolution is dominantly horotelic, and it is often faster, the higher the grade.

As a concrete example of this sort of evolutionary pattern, the Caenolestoidea, a superfamily of South American marsupials, may be cited (see Simpson 1939). There are here four very distinct grades of specialization: Caenolestinae, Palaeothentinae, and Abderitinae, within the Caenolestidae, and a fourth so advanced that it is separated as another family, Polydolopidae (see Fig. 21). All four appear fully formed and as if of equal antiquity in the record, but each grade could be and in all probability was derived from the next lower grade by a sudden shift in adaptive type. The Polydolopidae, most specialized or most narrowly adapted and least tolerant in ecological relationships, became extinct in the Eocene and the Abderitinae and Palaeothentinae disappeared during the Miocene. The Caenolestinae survive today and are essentially living Paleocene or Cretaceous mammals.

This example illustrates not only an evolutionary pattern so important that it will be discussed in more detail on later pages but also what appears to be a fundamental relationship more immediately pertinent in this chapter: liability to extinction tends to be directly proportional to rate of evolution. Bradytelic lines are almost immortal. The majority of tachytelic lines quickly become extinct and those that survive cease to be tachytelic. On an average, slow horotelic phyla live longer than fast horotelic phyla. Here reference is to absolute extinction, not to the conventional disappearance of genera by changing into other genera.

Rates of evolution tend to accelerate—not necessarily or usually within a given phylum, but in the sense that more specialized branch phyla tend to evolve more rapidly than their parent stock. A concomitant of this complex set of relationships is that younger branches often become extinct before older branches or before the parent stock. These are not the only factors involved in extinction, and this is not a "law of nature," but it happens often enough to indicate that it is a real and important part of the evolutionary pattern.

When related phyla die out in the order of their rates of evolution or in the reverse order of their times of origin, it follows that this order is also usually that of degrees of specialization and that more specialized phyla tend to become extinct before less specialized. This phenomenon is also far from universal, but it is so common that it does deserve recognition as a rule or principle in evolutionary studies: the rule of the survival of the relatively unspecialized. Application and exemplification of the rule, as well as identification of the numerous exceptions to it, require consideration of the concept of specialization, which is both complex and vague. Adequate discussion would lead far afield from this immediate topic, and for present purposes it suffices to consider specialization, not as distance from a defined primitive condition or postulated point of departure and not as intensity of adaptation, but as specificity of adaptation or inverse width of the zone of tolerance.

RELICTS

The word "relict" as usually applied in biogeography means, quite simply, "a form of life remaining within a smaller tract than before," yet this innocuous word is involved in an outstandingly great confusion of ideas and clash of theories. Other connotations have crept in, notably the implication that a relict is an older form of life than its more widely distributed allies or that it has evolved more slowly. At one extreme there are students (e.g., Rosa) who maintain that a narrowly distributed group is *ipso facto* a relict, and at the other extreme it is maintained (e.g., by Willis) that very local distribution is sufficient evidence that the group is not a relict and that, indeed, relicts are rare or nonexistent. More moderate zoogeographers hold that relatively restricted area characterizes both the earliest and the latest stages of phyletic history and that it indicates a relict only in the latter (see Simpson 1940).

Difficulties of definition stem from Handlirsch's classic study (1909). He distinguished the following sorts of "relicts":

1. Numerical relicts. Groups once abundant and now rare.
2. Geographic relicts. Groups once widespread and now restricted in areal distribution.
3. Phylogenetic relicts. Groups surviving from remote times with little change.

Handlirsch gave examples of a logical fourth category of his system, but did not name and define it:

4. Taxonomic relicts. Groups once highly varied and now reduced to relatively few species.

The first two of these categories and also the fourth, here added to those of Handlirsch, evidently include related phenomena and tend to coincide, without doing so perfectly or in all cases. A group of limited range may yet be abundant where it does occur (e.g., *Limulus*) and a group of relatively expanded range may have few species (e.g., the subgenus *Felis* [*Puma*], now with approximately its greatest range, but with only one true species). Nevertheless, it is generally true that a contracting group becomes less not only in geographic range but also in numbers of individuals and numbers of species. Handlirsch's inclusion of phylogenetic survivals (or persistent groups, bradytelic lines, etc.) among "relicts" reflects a widespread idea that these, too, tend to coincide with his other categories, to be reduced in numbers and in area. Willis's incontinent attack on the whole idea of relicts (or relics, as he permissibly prefers to call them) has the unstated postulate that if geographic relicts exist they should be ancient, ancestral types, i.e., slow-rate lines. It is this supposed connection or possible confusion of geographic restriction and slow evolution that makes relicts pertinent to the present topic.

It is probably true, as Willis insists and few deny, that a new group of organisms, that is, one recently differentiated from a more primitive ancestry, tends to spread as widely as its adaptive tolerance and its ability to meet existing competitors will permit. But it is also true that after a greater or lesser period of expansion and wide distribution most groups tend to contract (Fig. 22). This phenomenon may not be universal, but it is so obvious in innumerable groups known as fossils that denial of its common occurrence is plain refusal to accept established facts. An old group is more likely to be in the contracting phase than a young group. An old group that has evolved slowly is by definition a "phylogenetic relict." Such a group is likely to be contracting, and therefore likely to be a geographic relict, because it is old, not because it has evolved slowly.

The rate of evolution apparently has no bearing on the matter, and demonstrably it is not true that geographic relicts are more likely to have evolved slowly than at moderate or rapid rates. Slow lines often tend to keep on expanding after their characteristic rate of evolution is established. Bradytelic groups of relatively late origin and probably now in their expanding or expanded stage are not as a rule recognizable as such, simply because their record is not long enough to prove that they are bradytelic. Horotelic and probably tachytelic lines also tend to contract after reaching an apogee and also eventually become relicts. In keeping with the abbreviation of their whole evolutionary pattern, rapidly evolving lines tend, indeed, to become relicts more quickly than do slower lines.

One has only to make at random a list of animals that are relicts by geographical definition to show that there is no association with rate of evolution. *Sphenodon,* which evolved slowly, horses and giraffes, which evolved at moderate rates, and elephants, which evolved rapidly, are all geographic relicts. It is also true that some notably slow lines have not become relicts or still occupy such great areas that their

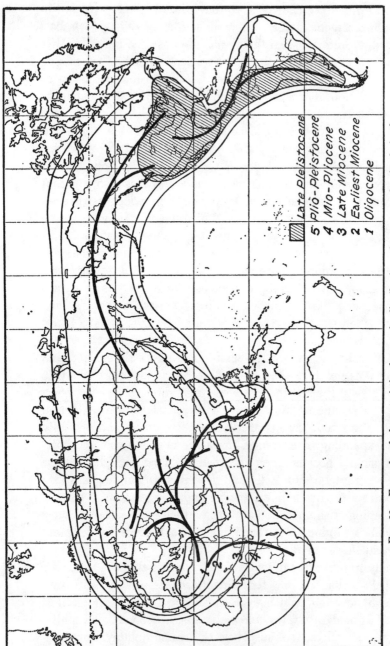

FIG. 22.—Example of characteristic changes in distribution of a group of animals in the whole course of its history. Approximate known geographic distribution of the mastodonts at the times indicated. The shaded area, occupied shortly before extinction of the group, corresponds with a time when the mastodonts were geographic relicts. Analogously expanding and contracting patterns are indicated for innumerable groups of animals and plants.

contraction is hardly enough to merit this designation. Cockroaches, crocodiles, and opossums are examples. The paleontological record of animals, at least, gives definite support to the conclusion that relicts are usually older than nonrelicts with comparable rates of evolution, but that they do not have slower rates than nonrelicts. It also suggests that for most animals the expanding phase is shorter than the contracting phase so that groups narrowly distributed are more likely to be relicts than new and expanding groups—a conclusion bitterly attacked by Willis, but on grounds little stronger than his personal disinclination to believe some of the best founded paleontological generalizations. Of course there are numerous exceptions, and the statement does not necessarily apply (as Willis seems to think it should) to groups that are merely fluctuating races or species adapted to localized conditions.

SUMMARY

The following are the principal theses that have been suggested or developed in this chapter.

Each group of animals has a standard evolutionary rate distribution, here called horotelic, with a strongly prominent mode and with frequencies of rates falling off steeply on each side, more steeply on the side of faster rates.

Different groups have markedly different absolute modal rates of evolution, but the form of the relative distribution of horotelic rates tends to be similar in all.

Within groups there are certain lines at certain times that have exceptional rates that fall outside the horotelic distribution of rates that are normal for most of the lines of the groups or for the whole group most of the time.

These exceptional rates may be slower than any in the horotelic range and may then be called "bradytelic," or they may be faster and may then be called "tachytelic."

Tachytelic lines either become extinct or usher in new major adaptive grades in which the phyla become on the whole horotelic, but often one line or a minority of lines are bradytelic.

It is not likely that a low rate of mutation or any one distinct category of individual characters is a decisive factor in bradytely.

BRADYTELIC SLOW
HOROTELIC MEDIUM
TACHYTELIC FAST

Bradytelic lines probably have continuously large breeding populations.

Bradytelic lines are not primitive when they first appear, but are then progressive and normally result from rapid evolution.

They often give rise to horotelic or tachytelic branches, but the bradytelic lines do not as a rule thereby become extinct as such.

The characters of bradytelic lines are predominantly and closely adaptive.

This adaptation is part of an equilibrium that involves the continuous accessibility of an environment or an ecological position which is either only slightly variable, with adaptation involving specificity of habit and low tolerance, or periodically variable, with adaptation to a corresponding range of tolerance.

In this condition of equilibrium almost all mutations are disadvantageous and tend to be eliminated by selection pressure as the controlling factor in a large population.

More recent and more rapidly evolving phyla are more liable to extinction.

Among a number of related phyla the less specialized, which are also likely to be the oldest and slow horotelic or bradytelic, often survive longer than the more specialized.

Most groups of animals tend first to expand and eventually to contract in geographic distribution; the rapidity of this development, but not its pattern, is influenced by rate of evolution, and relicts are not predominantly slowly evolving animals.

Chapter V: Inertia, Trend, and Momentum

IN THE ENORMOUS COMPLEX of observed or postulated evolutionary phenomena, even the listing of which would be a herculean task, certain sequences stand out because, disparate in other respects, they seem to have in common some element suggestive of the mechanical law of inertia. Bradytely is one such phenomenon: phyla evolving very slowly or not evolving except for small and apparently random variations on a constant theme sometimes continue indefinitely, even for hundreds of millions of years, in this evolutionary state of rest. It is doubtless true in evolution, as in mechanics, that acceleration may follow the application of something analogous to force, such as increase of mutation or of selection pressure, but even here the analogy, which is of a sort that must be accepted with great caution in any case,[1] begins to break down. Organic change is so nearly universal that a state of "evolutionary motion" is inherent in phyletic survival. It is probable that the continuous application of some sort of force, such as selection pressure, is necessary to maintain a state of rest and that the mere removal of restraint may be followed by acceleration. Moreover, unusually rapid evolution of many, if not of all, sorts tends to be self-limiting. It does not continue indefinitely until stopped by an external impediment, but quickly runs down if left to itself.

It would still be possible to say of these phenomena that, for instance, the braking of exceptionally high rates is a result of the exhaustion of the potential energy by which they are maintained, but such closer spinning of the analogy is no more than figurative description. It is not really explanatory, and it too easily leads to the assumption of analogy where none really exists. The evolutionary motion of a phylum is not at all motion in the same sense as is the motion of a mass in space. There is no theoretical expectation that the former will follow Newton's first law and no observational support for the hypothesis that it does so.

[1] Abel (1928, p. 97) supports in part the application of the laws of mechanical motion to the interpretation of evolution by the statements that living organisms necessarily function and that function is activity and consequently motion ("Funktion ist Taetigkeit und somit Bewegung"). This is naïve.

150 Inertia, Trend, and Momentum

The inapplicability of the basic mechanical law to evolutionary acceleration and deceleration, although it does not mean that these phenomena must have nonphysical determinants, casts doubt on any attempt to interpret related phenomena in the simple terms of classical mechanics. Few evolutionists do insist on the analogy in its application to rates, in themselves, but almost all agree that there are elements of evolutionary direction that resemble inertia and momentum, and it is a crucial, disputed problem whether this resemblance is only a descriptive analogy or whether it indicates some common causal elements and, in either case, how far the analogy is valid.

The principle of evolutionary inertia is that phyla tend to evolve in one direction without deviation. The related, but not necessarily consequent, principle of evolutionary momentum is that when a phylum is evolving in a given direction it tends to continue in that direction, even though the result may be disadvantageous or eventually lethal. Obviously neither principle, in itself, does more than state the supposed existence of certain phenomena, although both have been advanced under various guises or names as if they were explanatory. The important questions are: Do these phenomena exist? If so, how widely and under what conditions? Can they be correlated with primary or more nearly causal factors of evolution?

RECTILINEAR EVOLUTION

The evolutionary analogue of inertia is most widely known under the name "orthogenesis," proposed by Haacke (1893) and popularized by Eimer (1897) and others. Unfortunately the term is usually employed not simply as the name of a phenomenon but also as the designation of some theory purporting to explain it. Plate (e.g., 1913) proposed calling the phenomenon "orthevolution," but that term has been little used, perhaps because of its inelegant etymology. Plate supposed orthevolution to be due to two quite different causes, orthogenesis, redefined as the direct and rectilinear influence of the environment on the organisms concerned, and orthoselection, a term that is self-explanatory and has been rather widely employed, although hardly more respectable etymologically than "orthevolution." Although Osborn was probably the most intensive student of what most paleontologists call orthogenesis, he wished to avoid the causal implications given that word by Eimer, Plate, and others and spoke at first of "definite variation" (1889), then of "rectigradation" (1907), and finally of "aristogenesis" (1934).[2] For Osborn the pattern usually described as "orthogenetic" is the result of "a creative process from the geneplasma of entirely new germinal biomechanisms; the process is continuous, gradual, direct, definite in the direction of future adaptation." The development of Osborn's ideas and an important and useful series of factual examples of modes of evolution are given in the series of his papers cited in the Bibliography.

To add to the terminological confusion, many paleontologists and others confine the word "orthogenesis" to processes like aristogenesis, which involve some internal driving force, whether they believe in this or not (e.g., Lull 1921), a usage almost as different as possible from that of Plate and his followers. Still others seem to think of orthogenesis as any sort of evolution in which the direction is fixed by some force other than direct influence of or on the particular character under consideration (e.g., Weidenreich 1941), whatever the direction or force may be. Some students, avoiding the ambiguity of "orthogenesis" and the implication of "rectigradation" or "aristogenesis," also speak of "trends," for instance, in programme-evolution (e.g., Bulman 1933), but a trend is not necessarily rectilinear, which is the essential element of "orthevolution" or "orthogenesis." Gregory (e.g., 1935c) speaks more clearly of undeviating evolution. (See also the series of papers by Gregory, 1924 to 1937, for paleontological examples of modes of evolution and for more orthodox interpretations of some of the same sorts of phenomena stressed by Osborn.) Perhaps "rectilinear evolution" is the most purely descriptive available term, most nearly self-explanatory and least likely to evoke any particular theory as to its cause.

The innumerable and varied theories purporting to explain rectilinear evolution may be grouped in three broad schools, corresponding more or less closely with the major schools of evolutionists in general. (1) Direct interaction of organism and environment, regardless of the mutations that arise spontaneously or of the germinal modifications independent of the environment. This is orthogenesis in the restricted sense of Plate, and it is involved in the theories of many more or less

---

[2] As his ideas developed, Osborn used these and other terms for similar processes in various ways, and it is not true throughout his work that all apply to the same process or that this process is exactly what others call "orthogenesis," but at least once (1934, p. 209) Osborn did plainly say that definite variation, rectigradation, and aristogenesis are the same.

neo-Lamarckian paleontologists (Cope, Abel, and others). (2) The effect of natural selection on the survival and distribution of spontaneous mutations. This is Plate's "orthoselection" and the "orthogenesis" of most neo-Darwinians (Haldane, Wright, and others). (3) The occurrence of definite modifications of direction without reference to the environment, unless in the sense of preadaptation or predestination. This embraces a particularly wide range of theories, many of which are strongly antagonistic to the rest and have in common only the idea that orthogenesis indicates either an inherent trend or a more or less metaphysical tendency for evolution to proceed in straight lines. For some this is nothing more than a directional tendency in ordinary genetic mutation, a modification of or element in preadaptation (Cuénot and others). Most theories of this school, however, involve an element of predestination, of a goal, a perfecting principle, whether as a vitalistic urge, or a metaphysical necessity, or a frankly theological explanation of evolution according to which it is under divine or otherwise spiritual guidance (Osborn,[3] Vialleton, Teilhard, Broom, and a host of others).

There is no possible doubt but that some degree of rectilinearity is common in evolution. It can seldom or never be maintained that the evolution of a given phylum is exactly rectilinear or literally undeviating, but the best part of the paleontological record is made up of lines that evolve approximately in one direction over long periods of time. It is, however, obvious that rectilinear evolution is far from universal. The organisms that do exist and that have existed are completely inexplicable unless change of evolutionary direction has very frequently occurred not only by the divergence of branching phyla but also within single phyla. It is doubtful whether an undeviating or even a relatively straight structural line can be traced from an archetypal protozoan to any real metazoan, an ancestral fish to any real tetrapod, a protolemur to any existing primate, and so forth. The major changes of direction are systematically poorly represented in the fossil record, a point already stressed, but there are numerous examples of changes and even of complete reversals in the direction of evolution of organ-

---

[3] Although it is highly philosophical, Osborn's theory was definitely not theological, and he denied that it was metaphysical or involved either predetermination or a perfecting principle. This is a matter of definition. Aristogenesis is "definite in the direction of *future* adaptation," and the word means "to bring into being the *best* of its kind" [italics mine].

isms on a minor scale or of single structures, for example, the secondary simplification of ammonite sutures, the reduction of the canines in the Felinae, or the dwarfing of certain races of elephants.[4]

Although inertia, rectilinearity, orthevolution, orthogenesis, or whatever one wishes to call it, certainly occurs, it appears to be most characteristic of one part of the fantastically intricate pattern of evolution in general, and not necessarily the most important part. It does also occur elsewhere, and it may appear anywhere for short stretches of geological time (any curved line appears nearly straight if only a short segment is seen); but it seems to be a dominant pattern only on middle levels corresponding more or less with generic and tribal, perhaps subfamily, lines in classifications.[5] Moreover, the conclusive examples of observed rectilinearity are almost without exception drawn from groups with large populations evolving at moderate rates (probably always horotelic) radiating on low levels within a defined ecological sphere or following one ecological zone. It may be said against this conclusion that such conditions are the only ones for which many long and good fossil records are available and that the rarity or nonoccurrence of clearly rectilinear series under different conditions is merely the result of deficient records. But, in the first place, there is no a priori expectation that the pattern of evolution typical for such conditions would be common to all situations—on the contrary, there is reason to think that it would not; and, in the second place, extrapolation based on this assumption is frequently inconsistent with the facts.

The probability that a tendency toward rectilinearity is not characteristic of evolution as a whole, but only of certain levels of change under certain common but far from universal conditions, is in itself a potent argument against the third school of orthogenesis, that involving an inherently directional process or a metaphysical perfecting principle.

There are at least two directional genetic factors which cannot be doubted. They explain much of the linearity of evolution, but they

[4] Every example of the sort can be, and most examples have been, denied by students who say that *because* evolution is orthogenetic, the apparent ancestors of these forms are not the real ancestors and that therefore reversal did not take place. Such circular argument can be used to prove anything and has no place in an investigation trying without prejudice to infer what did happen.

[5] Naturally this varies with different students, and some students almost always equate an observed or inferred rectilinear sequence with a certain taxonomic category, such as the subfamily, simply because it is such a sequence.

alone cannot explain it all. They tend to weaken rather than to strengthen the hypothesis of inherently directional variation, because they explain much of the evidence for that hypothesis without recourse to the factors exclusive to it. First, heredity is mainly conservative, a fact so obvious that it seems to require no attention and is often overlooked in concentrating on the more puzzling and dynamic fact of change in heredity. In groups that do show rectilinearity, if not in all groups whatsoever, every animal is nearly like its parents in thousands of characters and may differ significantly only in a few characters that are really concerned in the long-term progression of the group. Second, it is not supposed by even the most rabid neo-Darwinian that there is nothing directional in mutation. It is not only improbable but also inconceivable that mutations in every imaginable direction occur with equal frequency. Some definite limitation of possible trends is inherent in the mechanism of mutation. But this is not to say that mutations often, or ever, occur in one direction only, that they tend to coincide with racial advantage or progress, or that this negative inherent limitation is a positive inherent directing force.

The conservative factor of heredity greatly limits the possible avenues of evolution for any given type of organism. If an animal is like its ancestors in all but a few respects, the differences must necessarily exist in connection and in harmony with the more extensive, more complex inherited elements of structure. "Granted that a character is dependent on the interaction of many genes, it will be easier to continue a line of evolutionary change, for which many of the modifiers are already present, than to start off on a completely new line. Thus the uni-directional nature of trend evolution is not particularly surprising" (Waddington 1939, p. 296). The fact that species are not constructed *de novo*, but on the basis of genotypes already existing strongly limits the possible avenues of change, and the possibilities are still further limited by natural selection, the restricting influence of which is obvious and is admitted by all theorists to some extent. Most often it would happen that no sort of available modification would be definitely advantageous and next most frequently that only one would be of selective value. In this situation a character would usually tend not to change or to change only in one direction.

The most crucial possible test of the probability of selection versus internal control of rectilinearity or the relative contribution of these

Inertia, Trend, and Momentum

two possible factors is to determine whether or not mutations (in the broadest sense) are directional and parallel to the actual direction of evolution. There is abundant evidence that the possible directions of mutations are limited. Most of them are of the "more or less" type, inherently limited to the two directions of a straight line, and the whole range of conceivable modifications in structure is never found to be covered by definitely observed mutations. Moreover, different genes mutate at characteristic and decidedly different rates, different mutations at one locus appear at different rates, and forward and backward mutations may have quite distinct frequencies (almost any study of mutation rates, for instance, Demerec 1933 and 1937, and Timofeeff-Ressovsky 1933).

Here is a possible mechanism for directional evolution under internal control, without reference to environment or selection, but other aspects of the same experimental data not only negative their potential importance for this question but also suggest that rectilinear evolution occurs in spite of directional mutation, not because of it. The observed directional effects are not such as do or, in many cases, could characterize observed and inferred rectilinear series. For instance, in *Drosophila* Timofeeff-Ressovsky (1933) found that the mutation of eye color to white is much the most frequent, occurring about twice as often as all other directions of mutation combined, not only from the wild type but also from all other alleles. Back mutation from white to the wild type (red) was not observed at all, and the only back mutation to the wild type observed (from eosin) was exceptional (two observed in 69 mutations of this gene). Here is strongly directional mutation, but it clearly is not in the direction of the actual evolution of any wild population.

For various reasons this particular example can be considered oversimple and rather crude in relation to the principal known rectilinear series, but similar considerations apply to all the observational evidence for directional mutation. Although many established mutations have a predominant direction and hence cannot be called fully random in this sense, it is clear that the favored direction is random with respect either to selective factors or to the direction that evolution of the group has followed. With curious logic the fact that most known mutations are not in the direction favored by evolution has been used as an argument against selective control of evolution or of orthogenesis, but in fact this is exactly what must be expected under that theory. If the actual trend and current adaptive condition are in the main controlled by selection, then most of the mutations occurring at any one time must be inadaptive if they are random relative to the rectilinear direction.

All this evidence decidedly opposes primarily genetic control of direction of evolution, and it is conclusively opposed to theories involving an internal perfecting principle. There is considerable evidence that if internal causes, such as directional mutation, do become effective in rectilinear evolution, the result is degenerative rather than progressive. For instance, Fenton (1935) concludes that "in *Atrypa*, at least, only degenerative (genic?) variations seem to be orthogenetic" and adds that in this genus, a slowly evolving Paleozoic brachiopod, "conditions seem to have been very close to those presented by the living *Drosophila*."

Furthermore, it is noteworthy that rectilinear evolution is characteristic of, if not quite confined to, just those conditions under which mutation pressure would be least effective in determining either the direction or the rate of evolution according to the theories of population genetics developed by Wright and others. Rectilinearity by directional mutation should be typical of very small populations under relaxed competition and slight selective pressure. There is, however, little evidence that markedly rectilinear evolution ever occurs under such conditions and every reason to believe that they are the least favorable to it. Such evolution would almost certainly lead to rapid degeneration and extinction.

Another criterion widely used in judging the probability of these opposing theories is based on the fact that new characters usually appear very gradually in more or less rectilinear series. A horn has selective value when well developed, but will an increase in its size be favored by selection when it is barely incipient? A complex ammonite suture has a conceivable advantage over a simple suture, but in the gradual change is a barely perceptible increase in complication of sufficient value to make selection favor continued change in that direction? Some degree of mimicry is evidently adaptively useful, but do predators really distinguish between approximate resemblance and the very exact resemblance often involved in mimicry? Such examples, which are very numerous, are standard items of evidence in favor of an inherent factor of some sort in orthogenesis, independent of selection, for instance, by Osborn (almost all his evolutionary studies), Robson and Richards (1936), Willis (1940), and many others.

This is a forceful argument against orthoselection; it involves, indeed, the only evidence against that theory that seems to me to be of much value in this sense. An observational approach to the problem has proved to be unusually difficult, and the reasonable but subjective opinion of one author that a slight variation or incipient character can have little selective value has been countered by the theoretical demonstration by another author that exceedingly slight selective pressure may nevertheless be influential in just such populations as do show orthogenesis (e.g., Fisher 1930). Recently experimental and observational evidence has also been accumulating that, despite the errors made by enthusiastic early Darwinians, slight modifications do really have a differential effect on mortality (e.g., Sumner 1935; Cott 1940). Without supposing that the question is definitely settled, it can at least be said that the observed orthogenetic phenomena of this sort are not inconsistent with orthoselection (see also Chapter II).

Theories of orthogenesis, strictly speaking, by direct environmental action are consistent with many of the same observational data as theories of orthoselection. If it is true, as it seems to be, that rectilinearity is most common at certain levels of evolution and in certain types of populations, orthoselection offers an acceptable explanation for this limitation, and orthogenesis due to direct influence of the environment does not. This is admittedly somewhat vague and inconclusive. The crucial criterion, again, is the old problem, whether acquired or externally induced variations can be inherited, and, again, I do not propose to discuss this problem, but only to state the opinion that this hypothesis is unnecessary, inadequately supported, and improbable.

EVOLUTIONARY TRENDS IN THE EQUIDAE

The most widely cited example of orthogenesis, in any sense of the word, is the evolution of the horse, and it has been analyzed from this point of view in considerable detail, notably by Abel (1928 and 1929) as exemplifying an evolutionary "law of inertia" (Trägheitsgesetz). Almost all such discussions seem, however, to be misleading as to what orthogenesis is as a phenomenon if it be defined as the sort of linearity seen in equid evolution, or, conversely, to give a mistaken impression of the extent of orthogenesis involved in equid evolution if orthogenesis be defined in some more theoretical way. No time need be wasted on the idea, conveyed by most of the popularizers of this topic, that equid evolution was the transformation of *"Eohippus"* (*Hyracotherium*) into *Equus* by a continuous, perfectly correlated, harmonious remodeling process going forward at a constant rate. Statements only slightly less naïve are, however, common in the technical literature and are seriously advanced as evidence for various theories of evolution.

It must again be emphasized that the study of equid evolution has barely been begun. Early students were concerned simply with proving that such evolution did occur, that our horse is a descendant of the little *Eohippus*. More recent students have been mostly interested in tracing the phylogeny and descriptive morphogeny and in establishing a taxonomy particularly useful for stratigraphic purposes. Study of the evolutionary principles involved has been almost wholly superficial and subjective. Many of the simplest data necessary for such a study have not even been compiled. Therefore much of what can now be said is still superficial, and much consists only of opinion as to what these data are most likely to reveal when they are compiled.

It was shown previously in this study that rates of evolution were not constant in the Equidae, but varied markedly from character to character, from time to time, and from phylum to phylum. Inertia as a tendency to retain a constant rate of evolution here operates, if at all, only within broad limits, and such a tendency is not established as part of the phenomenon of rectilinear evolution, although sometimes it is so considered. It is well known to paleontologists, apparently less so to other zoologists, that the Equidae show considerable phyletic branching, with at least twelve branches, aside from the direct *Equus* ancestry, on generic and higher levels and almost innumerable subgeneric and specific branches. An isolated, purely linear pattern is not typical of orthogenesis, or if orthogenesis be defined as involving such a pattern, the Equidae are definitely not orthogenetic.

Despite the great difference between *Hyracotherium* and *Equus*, most of the characters of the Equidae did not change appreciably throughout their history. *Hyracotherium* was already a vertebrate, a mammal, a placental, an ungulate, a perissodactyl, a hippomorph, and

Inertia, Trend, and Momentum

an equid, which is a classificatory way of saying that the vast majority of its multitude of morphological characters were already the same as those preserved in *Equus* and in all equids as well as in many other more or less related animals.

Of the characters that do show appreciable, evidently hereditary differences within the Equidae, many, probably the majority, do not appear to have any constant tendency to evolve in one and only one direction, i.e., so that their evolution is strictly rectilinear, if at all. If such a tendency affected all characters, except for relatively short periods of time, branching could occur only by rate differentiation. Rate differentiation is an important element in equid branching, but by no means the only element. Even within generic phyla and lesser groups for shorter periods of time, most characters do not have any parallel tendencies more marked than must necessarily arise in genetic systems similar in origin. They fluctuate, diverge, and even become reversed with what seems quite clearly to be response (in a selective sense) to varying and localized ecological conditions and is certainly not orthogenesis in any usual definition.

There does not seem to be even one distinct character that evolved continuously in one direction among all the Equidae. Some characters apparently evolved only in one direction (molarization of the premolars?), but not continuously; others may have evolved continuously, or nearly so (skull or limb proportions?), but in more than one direction. The simple picture that is usually evoked by the term "orthogenesis" is not realized here except as an average or statistical mode among lines with many differences and irregularities of trend. Among the more important or more obvious characters that do seem to show some modal tendency of this sort are the following:

1. *Gross size.*—There is clearly a tendency toward larger size in the Equidae, as there is in the great majority of vertebrate groups. Gradual increase in size is one of the most obviously advantageous specializations under most, but not all, conditions. The tendency in the Equidae is not universal and fluctuates irregularly, as if in adaptive response, not as if due to a directional genetic trend. At all times the horses covered a considerable range in size. Known species of *Orohippus* (middle Eocene) happen to average slightly *smaller* than known species of *Hyracotherium* (early Eocene). The difference is not statistically significant, but it shows that full continuity of trend toward larger size

is subjective and not supported by the data. The diagrams of steady increase of size are made by selecting species that fit this preconceived idea. Moreover, in at least two branch phyla, *Archaeohippus* and *Nannipus*, the trend was reversed; members of these genera are decidedly smaller than their probable ancestors.

2. *Skull proportions,"preoptic dominance," etc.*—These proportions show close simple heterogony with gross size, and there is as yet no evidence that this growth relationship changed or evolved within the Equidae, being apparently the same in *Equus* as in *Hyracotherium* (see pp. 4 ff.).

3. *Brain.*—Between *Hyracotherium* and *Mesohippus* the equid brain was remarkably transformed at an accelerated rate. Thereafter this trend and this rate did not continue, and there was relatively little change, mostly in correlation with size. Possible peculiarities of the dwarfed and other branch phyla are not known (Edinger, unpublished reports, 1941 and 1942). A broad trend toward greater intelligence, disregarding fluctuating details of brain structure, is obviously of selective value to animals such as the horses under any ecological conditions.

4. *Limb proportions.*—Although these characters have been much studied, the data on them have not been adequately analyzed. The average tendency was toward differential elongation of distal limb segments. It is not clear whether this growth pattern was present in early equids or evolved within the group. Such evolution as did occur was manifestly adaptive in general character, tending on the average to promote more efficient and rapid locomotion, and it was somewhat irregular, some specific and generic phyla showing probable reversal and others probable acceleration of the average trend.

5. *Foot mechanism.*—This crucial part of equid evolution had been almost wholly neglected until recently, except for a few conjectures of dubious value; but now there is a remarkable pioneering study by Camp and Smith (1942). They show that from *Hyracotherium* to *Mesohippus* the foot was supported by a pad and that between *Mesohippus* and *Merychippus* the pad was lost and a spring or recoil mechanism was developed operating on the hoof as a fulcrum. The spring mechanism persists in all later forms, both three- and one-toed, with various changes in detail and apparently also with divergent develop-

ments in the persistently three-toed and the progressively single-toed lines, although the evidence on this point is not yet wholly clear. From *Hyracotherium* to *Merychippus* the ligaments concerned in this evolution increase in number, size, and complexity. From *Merychippus* to *Equus* certain of the ligaments increased in size, but others were reduced, so that the net effect is one of simplification. Throughout the history there is a trend toward increase in length and total bulk of the ligaments, a trend such as is usually called "orthogenetic," although until more than a few typical steps have been examined its true rectilinearity cannot be strongly asserted. With increasing weight and speed the load on the ligaments increased, and this trend is not only of obvious selective value but also a mechanical necessity in animals that did increase in size and that did rely mainly on flight for defense. In details the sequence is not rectilinear. The change from an evolving pad-foot to an evolving spring-foot is a distinct shift in the direction of evolution. The change from complication to simplification of the spring mechanism and for certain individual ligaments from increase in relative size to decrease can fairly be described as reversals of direction.[6]

6. *Reduction of digits.*—During the three-toed phase most equids, possibly all, had simple heterogony of lateral toes with size and elongation of the middle toe, digital reduction not being an independently evolving character (see p. 6). This correlated or secondary trend continued without evident change in some phyla, even in animals as large and on the whole as specialized as *Equus*. But in the *Merychippus-Pliohippus* transition, and only here as far as is known, a different growth pattern evolved, resulting in the loss of functional lateral toes. It is entirely superficial and is erroneous from a more analytical and genetic point of view, to consider this as a continuation of the previous trend, or as parallel to the trend in other contemporaneous phyla. It does not continue the heterogonic trend otherwise common, if not universal, in equids, but is a new character, a definite change in direction of evolution, an exception to, and not an example of, orthogenesis. If, let us say, *Hipparion* had survived and had become our horse instead of *Equus*, this would have been obvious and the three-toed heterogony

[6] Camp and Smith are not responsible for stating the interpretation of trends in just this way, but their data and interpretations are consistent with this statement.

would have been recognized as the more rectilinear of the two trends. It is only the accident that *Equus* did survive that makes us think of its ancestry as central or typical.

7. *Molarization of premolars.*—This was not a continuing trend of the Equidae, but occurred entirely within the essentially unified ancestral phylum *Hyracotherium-Mesohippus*. All known divergent generic or higher branches are later than and derived from *Mesohippus*, and they simply inherited molariform premolars without further evolutionary advance in this respect. After *Mesohippus* the premolars evolved in unison with the molars, but they did not become any more like the molars. This unification of the cheek-tooth battery is adaptively advantageous for the whole range of possible equid food habits because it increases the efficiency and speed of mastication.[7] Subsequent evolution of premolars and molars (nearly, but not absolutely) in unison may have resulted from the selective advantage of maintaining unity of the dental battery, or by its being brought under the same genetically controlled developmental field, or by both.

8. *Hypsodonty and cement.*—As previously remarked (pp. 7 ff.), hypsodonty is not heterogonic in equids. Through *Parahippus* and in the later or parallel *Anchitherium, Hypohippus,* and *Archaeohippus* phyla, the relative height of crown increased slightly in approximate proportion to its size and in apparent adaptive response (such as would be ensured by selection) to the fact that larger animals need more durable teeth for the same diet. In *Merychippus* true hypsodonty and cement (a wholly new character) appeared. This was not a continuation of the *Hyracotherium-Parahippus* trend; it was a definitely different direction of evolution. Again, if *Hypohippus* were our living horse, it would be obvious that *Hyracotherium-Hypohippus* is the rectilinear line in this respect, and *Merychippus-Equus* a branch in a totally new direction.

9. *Molar pattern.*—Earlier perfection of lophiodonty and later complication of the lophiodont pattern are parts of the trend probably most widespread in equids, but again it is not such as would be expected if rectilinear evolution were the result of determinate internal directional factors. Both lophiodonty and complication (at different

---

[7] A similar end is served in quite a different way in ruminants, which do not need to chew the food thoroughly while feeding in exposed locations, but complete the process subsequently in more protected places. In them the premolars do not become as molariform as in perissodactyls, and they may be more reduced.

times) evolved to what appears to be an average optimum and thereafter fluctuated locally around this modal adaptive type, without clearcut average progression. In one line, *Hypohippus-Megahippus*, there may even have been reversal toward some secondary simplification. The complication is achieved by a number of different elements that did not always evolve together or appear in all lines. Some parallelism of origin of such elements may be demonstrable, but the inadequate data now available suggest that the usual process was for a given element, such as the crochet, to appear as a variation in a population, slowly to spread to the whole population, and to be inherited by derivatives of that group, independent phyla developing different variants serving the general adaptive purpose of complication in the grinding surface.

This general picture of horse evolution is very different from most current ideas of orthogenesis. It may be agreed that there are rectilinear elements involved, but they are certainly less widespread and less persistent than is usually asserted for this classical example of orthogenesis. They are thoroughly consistent with orthoselection, which indeed seems the only reasonable explanation of these trends in the horses. They are flatly inconsistent with the idea of any inherent rectilinearity, predetermined trend or solely preadaptive control.

Opponents of the theory of orthoselection (or of selection in general) impute to that theory the requirement that every environmental change should produce a change in adaptation and that every direction of change should be toward the best possible adaptation to one precise environment. For instance, Willis (1940) has, to his own satisfaction, built up and then destroyed a straw man composed of assumptions like these and a great many others that not only are not real parts of the modern theory of natural selection but also are definitely rejected by the adherents of that theory. Abel (1928) has the Equidae not merely exemplifying but also providing the test case for evolutionary inertia because, he says, their direction of evolution remained the same though the environment varied from forest, savannah, and veldt to tundra, jungle, and desert. In the first place, this is not really true. There were decidedly divergent directions of evolution between, for instance, forest and desert horses. In the second place, such well-defined (never universal) trends as did really occur were quite clearly **advantageous** under *all* the varying conditions encountered by any equid phyla that show these trends. Not all conceivably useful trends did occur, and the observed changes are not always exactly those that to a mechanical engineer might appear the most advantageous possible, but no one seriously believes that selection must, or in fact can, act in so ideal a fashion.

A dispassionate survey of many of the phenomena of orthogenesis, so called, strongly suggests that much of the rectilinearity of evolution is a product rather of the tendency of the minds of scientists to move in straight lines than of a tendency for nature to do so. Gregory (1936c) remarked in closing an excellent summary of the relationships of "undeviating evolution" (orthogenesis in a descriptive sense) and "transformation" (change in direction of evolution): "Some future historian may note that toward the end of the first third of the twentieth century the very abundance of the new discoveries of fossil mammals resulted in obscuring temporarily the grand fact of transformation" (Gregory 1935c).

PRIMARY AND SECONDARY TRENDS

Since the earliest days of paleontology it has been noticed that some phenotypic characters tend to vary together, or be associated, or show functional correlation—three different phenomena that are difficult to distinguish. This is involved in the so-called Cuvierian principle of correlation, which even antedated Cuvier. For about a century no marked progress was made in understanding or formalizing these relationships beyond a certain disillusionment with the rigid and crude application of the Cuvierian principle caused by such things as the association of claws with perissodactyl teeth in chalicotheres in rank defiance of this man-made "law." Only within the last few years have more exact data been gathered and studied from a more analytical point of view.

In its most general and most purely objective aspect, the problem is that of simultaneous trends, a simple matter of observing that while one character changes in a certain way another character also changes in its way. It is obvious that this descriptive statement may cover quite distinct evolutionary processes, and little study is required to show that it certainly does so. Simultaneous trends may occur in two or more characters because (1) both are influenced by some one genetic factor, for instance, gross size and skull proportions in Equidae, (2) they are functionally related in the economy of the organism, for instance, size and crown height in brachyodont Equidae, or (3) there is an extrinsic coincidence of trends not necessarily subject to internal co-ordination, for instance, digital reduction and tooth complication in Equidae. Many of the Cuvierian correlations were products of the third sort of simultaneous trend. The lack of control by organism co-ordination makes possible the association of quite different trends in different phyla, and this is the reason for the failures of the Cuvierian principle.

Each of these three categories is found still to be multiple when closer analysis of more nearly causal factors is attempted. Characters with genetic correlation, definite covariates, may have simultaneous trends (1) because they fall within a unitary heterogonic or harmonious growth field, (2) because both are influenced by a pleiotropic gene, or (3) because of linkage of genes. Functional correlation may involve (1) direct coaction or interaction (as in occlusion of teeth), or (2) the necessity for distinct organs to perform congruent functions (as in the relationship of dental and digestive systems). Extrinsic coincidence may reflect (1) ecological correlation of separate characters, both advantageous in a given situation (such as adaptations for grazing and for rapid locomotion), or (2) accidental simultaneity (perhaps like the development of the mane and of self-coloring in lions, but such is the integration of structure and so pervading is the co-ordination of morphology and ecology that it is hazardous to assume that any simultaneous phenomena are independent). The various factors are not always absolutely distinguishable, and two or more may be involved in any one instance of simultaneous trends.

Extrinsic coincidence does not clearly involve an element of dependence of one trend on another, but in both genetic and functional correlation there is dependence, and it is natural to think in some such cases, if not all, that one trend is primary and independent and another, induced by or following this, secondary and dependent. This concept both simplifies and complicates the problems of evolutionary processes and morphogenesis—simplifies them by reducing the study of separate trends to the determination of a single set of correlating factors common to all, and complicates them by making it difficult to distinguish primary from secondary trends and to establish the nature of their relationship.

In an unusually able and stimulating study of the relationship between brain size and skull structure in man and other mammals Weidenreich (1941) has recently suggested that morphological characters can be grouped in three classes.

1. Fundamental or first-order characters.—These are characters common to all members of a family or higher group. The great weight of the brain in man is given as an example.

2. Special or second-order characters.—These may be generic, specific, or racial, and their distinguishing feature is that they depend on other morphological factors. An example is seen in the special characteristics of the skull in the various genera and species of apes and men, characteristics shown to be correlated with relative brain size.

3. Secondary or third-order characters.—These are specific, subspecific, and racial (and by implication are independent of the "fundamental" or "special" characters). Stated examples are hair texture and skin color in man.

Weidenreich believes that only "first-order" characters are essential to the problem of evolution, a problem that cannot be studied by genetics, because genetical experiments deal only with "third-order" characters. From the dynamic point of view, primary trend is shown only by the "fundamental" characters and the "special" characters follow these automatically, orthogenetically. Presumably the third-order characters are considered sporadic and not involved in definite trends.

Useful as it is, this classification has the failing that it establishes categories by multiple criteria that are only assumed to be coincident, an assumption that is, in my opinion, demonstrably false.[8] There is no reasonable doubt that not only the same sort of characters but also identically the same characters may in some cases distinguish "good" (clear-cut, broad, conservative, well-founded) families or genera and in other cases species or races. From this point of view "third-order" characters are not qualitatively different from "first-order" characters and they regularly become first-order characters by segregation, by

[8] A closely parallel taxonomic fallacy is developed by Goldschmidt, whose work is not cited by Weidenreich.

accumulation of small differences, or by both. Again, the distinction between correlated and noncorrelated characters does not coincide with the distinction between characters in common at different levels of classification.

The fundamental character that distinguishes men from apes and that has different skull characters as a corollary is not so much the larger brain, in itself, as the different growth pattern from which both a larger brain and a different skull result. An ape skull is very different from a normal human skull with a brain of the same size; if the heterogony formula were the same in men and apes, man would be an ape. This relationship is quite clearly demonstrated by Weidenreich, but the form in which it is generalized by his classification of morphological characters seems to me confusing.

In such situations, mainly dealing with characters genetically correlated by heterogony, but also seen in other sorts of correlation, the designation of primary and secondary trends may be difficult or impossible. Calling one of two correlated characters "primary" because it is considered more important has no evident validity unless there is some causal reason for its greater importance, particularly unless it is demonstrably a prior link in a concatenation. In the example of brain size and skull configuration there is at least a reasonable probability that the more nearly causal element is brain size on the theory that its increase was favored and, in this sense, produced by natural selection and that simultaneous skull changes did not have independent selective value. Without worrying too much about philosophical difficulties involved in causation and causality, it is fair enough to say that the skull changes were a result of increasing relative brain size. If, however, no particular reason for increased brain size were suggested, the opinion that it is the primary factor would merely reflect personal bias.

In many such cases not even this approach toward the causal connection implied by designation of trends as primary or secondary can be made. For instance, in differentiating between proportions of limb segments or length of limb relative to body size, it is merely misleading to say that the tibia became longer because the femur became shorter or that the limb became relatively elongate because the body became larger. Such characters evolve together, interdependent, not dependent-independent.[9] If the control is selection, this either acts simultaneously on both characters or acts, not on the characters as such, but on their relationship to each other.

A remarkably interesting suggestion has recently been made by Watson (1941), with the evolution of the Amphibia as its basis and example. He finds two primary trends in all amphibians, flattening of the skull and reduction of cartilage bone. It is concluded that these trends "are secular, they continue apparently uninfluenced by profound changes of the animal's habits, and must therefore arise from internal causes acting continuously throughout the whole period of existence of the group." Other changes in the amphibian skull, in some cases ultimately profound, are believed by Watson to be brought about by natural selection, maintaining functional co-ordination with the two primary trends and therefore, in a special sense, the result of the latter. These secondary changes differ in different lines, because hereditary variations differ; but there is much parallelism, because the repertory of suitable mutations is limited. Different lines also differ in the rates of the primary and the secondary trends, and, together with qualitative differences in secondary trends, these differences eventually made the various phyla radically unlike, although all were subject to the same evolutionary processes.

There is some question whether these primary trends are literally universal among amphibians, although a few possible exceptions do not alter the fact that they are widespread enough to require special explanation. Without particular reference to this example, such "universal" trends are sometimes artifacts and have quite a different meaning from that here implied. If a divergent trend becomes marked enough, the animals showing it are removed from the parent taxonomic group. For instance, the distinctly "nonamphibian" trends to reptiles, with relatively deep heads and relatively unreduced cartilage bone, did start among amphibians, and the absence of these trends in the Amphibia is an artifact of classification. The possible directions of other diverse changes are limited by the need for appropriate, coincident mutations. The unfavored trend may be immediately lethal without such rare coincidence, or distinct trends may become estab-

---

[9] Some confusion has arisen, for instance, in the discussion of the horse head proportions, because formal statistical treatment of the regression conventionalizes one variate as dependent. This has no bearing on biological or evolutionary dependence.

lished, but be at a relative disadvantage and soon disappear. For instance, a basic and all-but-universal trend in the Crocodilia involves depression of the skull and negative heterogony of snout height against length. It is known from a single specimen (*Sebecus* Simpson) that an opposite trend, with lateral compression and positive heterogony, did arise, but that it produced a relatively unsuccessful type of crocodile, never common and now long extinct. If this specimen had not been found, the crocodilian trends would be cited as evidence of "internal causes acting continuously," but the isolated discovery makes it obvious that this is not true. I have compiled, but have not published, a large body of data on these trends in Crocodilia. Of course, the discovery of exceptions does not prove that an internal cause does not exist, but the strongest argument for internal causes is the supposed absence of exceptions.

In the case of the Amphibia no selection value has been assigned with certainty to the two most general trends noted by Watson, but it is evident that such characters can have selection value and the assumption that they were developed by a known mechanism, mutation-selection, is at least as permissible as the assumption that they were the result of an unknown internal cause. A possible clue to selective influence may be found in the conclusion of Romer (1942) that cartilage is an embryonic adaptation and persists in adults because of abbreviation of the life cycle. In the more limited but clearer example of the Equidae it is sufficiently evident that the broader and more general trends that may persist regardless of changes in habits were simply adaptations effective in the widest range of circumstances rather than specific to one habitat or one habit. It may reasonably be suspected, even though it cannot be so clearly shown, that this is also true of the more widespread trends of the Amphibia.

Although the factors are different from those exemplified in the discussion of genetically correlated trends, in such examples as this there may also be primary and secondary trends. The primary trends, whether produced by wholly internal causes or by selection, may proceed without respect to others. The secondary trends may be advantageous in co-ordination with the primary trends, although not of selective value in themselves when unaccompanied by the primary trend. Here again, however, in many of these functional correlations it is impossible to designate one trend as definitely primary.

EVOLUTIONARY MOMENTUM

One of the most conspicuous contributions of paleontology to evolutionary theory has been the suggestion that phyla evolving in a straight line, in a direction originally adaptive, may acquire momentum that may carry them beyond the optimum and that may even cause their extinction. Unlike rectilinear evolution, which is a demonstrated phenomenon in a descriptive sense, the question whether there is such a momentum effect is still open. Whether a given structural sequence has passed its optimum cannot be conclusively determined unless all the factors concerned in fixing the optimum, physiological, ecological, and so forth, are determinable, which is seldom true. Opinion on this point is generally dubious and sometimes wholly subjective.

In view of these doubts, a few supposed examples must be specified. Bulman (1933) maintains that orthogenesis, in what he calls "programme-evolution," carried lobation so far in the monograptids that it caused their extinction. The phenomenon occurs in a line of graptolites in which the thecae, housing individuals of the colonial organism, became globular and recurved so that their apertures are near and directed toward the stem carrying them. Bulman is not explicit as to why such a development was harmful; presumably the thought is that extreme lobation led to lethal restriction of the aperture, a possible but perhaps not an established conclusion.

A similar and clearer example among invertebrates is provided by certain forms of *Gryphaea,* a pelecypod in which "in some individuals the umbo of the left valve actually pressed against the outer surface of the right valve, so that this could be opened only slightly if at all. Such a state of affairs could only lead to the death of the individual and the extinction of the race" (Swinnerton 1923). The progressive, descriptively orthogenetic development of this unusual condition is shown by the data for degrees of curvature at successive times in the *Ostra irregularis-Gryphaea incurva* phylum (Swinnerton 1932, in part after Trueman). The maximum curvature observed in five successive samples increased regularly from 130 degrees to 540 degrees, from less than half a coil to one full coil and a half. In representative individuals Swinnerton found the following interesting progression (oldest at top, youngest at bottom).

It is worthy of notice that this same phylum is also rectilinear in

TABLE 18
PROGRESSIVE CURVATURE IN THE SHELL FROM A PRIMITIVE *Ostrea* TO AN ADVANCED *Gryphaea*

Form	Degrees of Arching When Basal Lamellae Were 26 mm. in Length	Length of Lamellae at Onset of Gryphaeoid Arching	Degrees of Gryphaeoid Arching per Millimeter of Lamellar Length
Ostrea irregularis, primitive variant	20	26.0	0.6
Ostrea irregularis, typical	75	18.0	4.7
Gryphaea dumortieri	150	12.5	8.2
Gryphaea incurva, typical	210	2.9	8.6
Gryphaea incurva, advanced variant	250	0	9.2

most of its other observed characters; none of them seem to pass the point of viable adaptation and some of them seem to be increasingly advantageous throughout for a postulated mode of life (for extensive data see Trueman 1930 and earlier papers; also George 1933). For instance, the length of attachment relative to total adult length of scar decreased from about 50 percent in *Ostrea irregularis* to about 4 percent in *Gryphaea incurva,* and the proportionate duration of the fixed nepionic stage correspondingly decreased.

Cuénot (1925) has given numerous examples of this supposed momentum effect, which he calls hypertely, among them classic instances cited by almost every author who treats the subject: the elongation of the dorsal spines in the Permian reptiles *Edaphosaurus* and *Dimetrodon;* the length of the canines in the sabertooth *Smilodon,* and especially the hypertrophy of the antlers in the extinct giant Irish deer *Megoceros.* Haldane (1932) adds the great development of horns in the latest titanotheres as another example, and indeed there are few cases of extreme development of offensive, defensive, and apparently decorative structures, spines, horns, armor, and so forth, that have not been or could not be cited in this connection.

Yet these very examples show the impossibility of deciding categorically that a given degree of development is really disadvantageous. For instance, Cuénot weakens the case by citing structures equally extravagant, to our eyes, among animals that are not only alive but also thriving and that in some cases can be inferred to derive some advantage from their peculiarities, among them the brilliant displays of male pheasants, peacocks, and birds of paradise, the horns of numerous beetles, peculiar appendages in various living Hemiptera, and so forth. Note also babirussa tusks and the lower teeth of *Mesoplodon*, cited by Abel (1929.) Extinct animals present nothing more fantastic, and in the living animals these characters are at least not lethal.[10] Moreover, the many authors who speak of momentum effects in such characters as the machairodont canines overlook the fact that this character did not evolve steadily to a climax and then past it to extinction. The earliest known machairodonts in the early Oligocene had nearly (in some cases quite) as large canines, relatively, as did the terminal forms that became extinct millions of years later, in the Pleistocene. Far from overshooting the mark, the machairodonts seem rapidly to have reached it and to have retained constant adaptation, with the expected fluctuation, for a very long time. This is evidently not true, however, of more continuing progressive characters, such as the Irish deer antlers or the hypertrophied tusks of the late mammoths (an additional example stressed by Abel [1929] and others).

As Haldane (1932, p. 23) says, these phenomena "are not obviously explicable on any theory of evolution whatever," and Waddington (1939, p. 296) stresses that no explanation in terms of natural selection has yet been provided. Most students have been contented to describe what they considered momentum effects without attempting to explain them. Neo-Lamarckism most signally fails here, because it seems theoretically impossible for either the influence of the environment or the efforts of an organism to adjust to the environment to go beyond such adjustment to secondary maladjustment. The reality of the phenomenon would also contradict the theory of an internal perfecting principle. It would be in accord with the various theories of some sort of internal, directional drive and is usually cited as evidence for these theories, but they are far from explanatory, even in an elementary sense. They say, in effect, that the results of evolutionary momentum arise because evolution has momentum. This is not enlightening. It may be the best that can be done in the present state of

---

[10] Cuénot denies the sexual advantage of such characters as large antlers. He argues that if these were really useful weapons the males would have them all the time and the females would also have them, but it is well-known that antlers are used almost exclusively against other males in the breeding season and that, on an average, they are decisive in determining which shall be the fathers of next year's fawns.

knowledge, but I think not. There are several ways in which apparent (or, according to the definition adopted, real) momentum effects can arise within the scope of neo-Darwinian theories of mutation-selection in the organism-environment system.

A rather isolated attempt at this sort of explanation was advanced by Dendy a generation ago (1912). His suggestion was that selection for larger antlers, to use the example of the Irish deer, is selection against a factor inhibiting or limiting growth instead of, or in additon to, a positive factor promoting such growth. In such a case selection might conceivably operate so strongly that the secondary result, antler growth, overshot the mark. The suggestion is ingenious, and it has some inkling of processes that now appear more likely; but there is little evidence that selection does operate in such cases against a non-growth factor rather than in favor of a growth factor. If the inhibiting factor (Dendy suggests a hormone) were not wholly lost, reversed selection beyond the optimum would tend either to stop evolution at that point or to produce a rapidly damped oscillation about it. If, on the other hand, the factor were wholly eradicated, there would presumably be a tremendous saltation followed by immediate extinction.

A possibility that has considerable resemblance to Dendy's hypothesis is suggested by Fisher's work, although he has not quite stated it as such. If two different gene mutations tend to produce the same or similar advantageous results in the growth mechanism, both will be favored by selection, and as their gene ratios increase they may more frequently appear together in single individuals, which may produce an effect greater or less harmonious than the presence of only one. Similarly, it might happen that the heterozygote is advantageous, but the homozygote carries a given character beyond the optimum, and that selective favoring of the heterozygote would unduly increase the number of homozygotes. Both situations must, however, be somewhat unusual, and it is unlikely that either could carry the population as a whole decidedly beyond the optimum.

A suggestion made by Haldane seems to be more important and may be carried somewhat farther and applied to specific examples. Some sorts of selection pressure are most effective in the young. They may therefore favor characters that aid survival in the young, but are disadvantageous in the adult, either *per se* or as developed by the growth processes. Thus, in the example of the Irish deer, early development of large antlers, for instance, by a high degree of positive heterogony, may be advantageous to the young bucks and may be favored by selection among them, but may result in an excessive maximum antler size in older bucks.

It is often overlooked that hereditary factors that reach their expression only after adults cease to breed have little bearing on natural selection, which will not as a rule oppose them no matter how disadvantageous or lethal they may be or favor them no matter how useful. Thus, in *Gryphaea* the fatal blocking together of the two valves occurs only in the latest stages of life, to which only a fraction of the individuals survive in any case. They have already contributed their share, or more than their share, of progeny to the race before the heredity that they possess and transmit kills them. It is true of both this suggestion and of Haldane's related suggestion, as Waddington has said of the latter, that it is not clear that extinction of the species would result. But, in the first place, it is not clear that extinction did result from the supposed momentum effect in these cases; in the second place, both these factors would somewhat, however slightly, reduce total reproductive capacity in the population and could be an element in their extinction if environmental pressure were increased.

It is even theoretically possible in certain cases that hereditary factors limiting length of life and disadvantageous to late adult or senile individuals will be favored by natural selection because they lessen intraspecific competition and ensure that offspring will be those of young and vigorous individuals.[11] None of these processes leading to unfavorable characters late in life deserves to be called "evolutionary momentum," but they are fully competent and probable causes for some phenomena, notably in *Gryphaea* and analogous cases, that are usually called "momentum effects."

The nearest thing to a true momentum effect that is likely to occur in nature would result from coincidence in direction of selection pressure and mutation pressure. The optimum favored by selection would be passed if mutation pressure continued to be effective (Fig. 23A).

---

[11] One of the numerous fallacies in Willis's criticisms of natural selection is his repeated claim that natural selection acts only on individuals and should favor characters advantageous only to each separate animal. Natural selection acts, in part, by differential effects on individuals, but what is actually selected is not characters individually advantageous, but those racially advantageous. Such characters may even be definitely disadvantageous to a given individual. See, for instance, Haldane 1932, pp. 207–210.

For progressive adaptation to occur, one genetic locus or more must be mutating advantageously. If mutation rate is itself hereditary and variable, which is probable (e.g., Darlington 1939), selection may favor

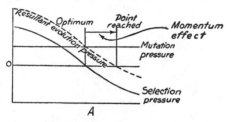

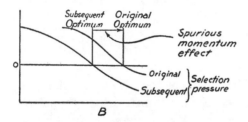

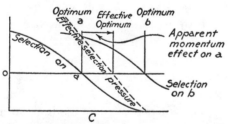

Fig. 23.—Analysis of some supposed momentum effects in evolution. The tendency of a given evolutionary factor is to carry the character to the point where the line graphing intensity of that factor intersects line *o*. *A*, Constant mutation pressure plus selection pressure about the optimum produces a resultant evolution pressure that intersects *o* beyond the selective optimum; *B*, backward shift of optimum makes original optimum inadaptive and produces spurious aspect of momentum for group that had reached the original optimum; *C*, correlation of two characters with different optima makes effective selection pressure different from selection for either character alone and places the effective optimum beyond the optimum for one of the characters.

indefinite increase in mutation at these loci, and such an influence would favor this phenomenon. Mutation pressure is, however, so ineffective in most populations, that notable progress beyond the optimum must be unusual and probably occurs only in small populations with slight selection pressure. In most cases, especially in groups best known as fossils, the effect would be absent or quickly damped out, and it is not a probable explanation for most of the supposed results of momentum in evolution.

What appears to me the most general explanation of these so-called examples has nothing to do with momentum. This is simply that a phylum has become highly and adaptively specialized and a change in selective influences has made the specialization secondarily inadaptive (Fig. 23B). In other words, the phylum approached or reached an optimum without passing it, and the optimum then shifted to a previous position, or to one outside the given line, faster than the population could follow, an aspect of what Darlington (1939) calls "lag," failure of the genotype to produce mutation as rapidly as selection requires. Indeed, this is probably the most general cause of extinction, and when the optimum shifts to an earlier point, the character may appear to have, but will not really have, progressed beyond the optimum.

This conclusion was clearly expressed by Matthew long ago, and his statements are as valid now as then, despite the fact that the machairodonts, which he was discussing, continue to be cited as examples of momentum by writers whose knowledge of them is incomparably inferior to Matthew's. He wrote (1910, p. 307) that advocates of the theory of extinction by momentum "have repeatedly quoted *Smilodon* as an example in support. In point of fact, as we have seen, the immense development of the canines in this animal made them highly efficient weapons for a particular mode of attack and was an essential element of its success in its especial mode of life, not a hindrance or bar to its survival. Whatever may be thought of the theory of 'momentum in evolution,' *Smilodon* cannot be used as an instance in support." It was an environmental change, due to the scarcity and final disappearance of the prey for which the machairodonts were specialized, that probably caused their extinction. Selection pressure was reversed more rapidly than machairodont mutation could follow: lag, but not momentum.

A final possibility is that the progress of a structure beyond its optimum really occurred, but that it was caused by genetic or heterogonic linkage of this structure with another that had not reached its

Inertia, Trend, and Momentum

optimum. Then the true selection value is not the expression of one alone, but the two together. The genetic optimum is reached when the disadvantageous development of one, which has passed its phenotypic optimum, just balances the advantageous development of the other, which has not reached its phenotypic optimum (Fig. 23C). Such a phenomenon provides an elegant and sufficient explanation for many supposed momentum effects, including another classic "proof," the Irish deer, in which the antlers became so large that they had probably passed the optimum size. If this group had strong positive heterogony of antler size against body size, as it seems to have had and as its ally *Cervus elaphus* demonstrably has (Huxley 1932), then the increase in body size automatically produced a disproportionate increase in antler size. The condition reached was that at which the disadvantage of further increase in antler size balanced the advantage of further body increase. Here, again, postulation of a true momentum effect is quite unnecessary. Forms of linkage other than that of relative growth rates could and doubtless do produce the same sort of phenomenon. Extinction would not follow automatically, because the point reached between the two phenotypic optima is the evolutionary optimum; but, again, there is no real evidence that the supposed momentum effect did cause extinction. It could readily be a contributing factor, because a group that suffers evil that good may come, that necessarily acquires a disadvantage to obtain an advantage, is likely to be less successful in any intensification of the struggle.

THEOREMS ON INERTIA IN EVOLUTION

The following are some of the more important propositions supported by the data and discussions in this chapter.

A tendency for phyla to continue to evolve in much the same direction for considerable periods of time, rectilinear evolution or orthogenesis in a purely descriptive sense, is a common evolutionary phenomenon.

Rectilinear evolution is not universal and is most typical of large populations evolving at moderate rates.

Rectilinear evolution does not involve simple, linear, unbranched phyletic patterns, and it may be accompanied by pronounced changes in rates of evolution.

Much of the linearity of evolution is inherent in the predominantly conservative effects of heredity.

Mutation is frequently directional in the sense that it occurs more frequently in one direction than in another, but it is usually random in the sense that this favored direction has no special tendency to coincide with advantageous modification or with the direction in which the group is really evolving.

Control of evolution by directional mutation is usually of short duration, except possibly in small degenerating groups.

Progressive rectilinearity in evolution may occur rather in spite of favored mutational directions than because of them, although there must be some mutation in the evolutionary direction, and this may be increased by natural selection.

Continuity and wide distribution of linear trends are usually in inverse proportion to their specificity from an adaptive point of view, trends advantageous under the widest variety of circumstances being most enduring, most general, and most rectilinear.

Observed rectilinear sequences are usually most consistent with the theory that orthoselection is a dominant or primary factor, and some are not logically explicable in any other way.

Various genetic, physiological, and ecological factors may produce secondary trends, not directly and immediately orthoselective, but correlated with a selective trend.

There is no good evidence that a trend has ever continued by momentum beyond a point of advantageous or selectively neutral modification or has ever been the direct cause of extinction.

A real momentum effect could result from mutation predominantly in one direction, but only under unusual conditions; this has not been an important factor in evolution, and it is not definitely known that it has ever occurred.

Apparent momentum effects may be produced under the influence of selection in a variety of ways; among them are: selection on juveniles or young adults favoring characters disadvantageous to old adults; shifting of the optimum to a previous point faster than the population's genetic structure can follow; simultaneous selection on correlated characters such that the optimum for the two together is reached after one has passed its separate optimum.

Such effects can hardly be primary, but they may be secondary causes of extinction to the extent that all tend to reduce the capacity of the population to react to changing conditions.

Response to selection pressure is not instantaneous, and inertia, in the sense of lag in following a shifting optimum, is an important element in evolution.

Chapter VI: Organism and Environment

THE ASPECTS OF TEMPO AND MODE that have now been discussed give little support to the extreme dictum that all evolution is primarily adaptive; but they show how necessary it is to consider evolution in terms of interaction in organism and environment. In this system adaptation is the crucial element. Whether adaptation be considered a cause, an effect, a concomitant, or a goal may not really be of primary importance for its study. Often these disputed points are matters rather of viewpoint or of terminology than of real distinctions. Adaptation may be any or all of these things, but it exists, it is universal in some degree, and it arises and changes in the course of the genetic and somatic modifications that are evolution.

Before proceeding to the final aim of this work, an analysis of the principal elements in the pattern of evolution (Chapter VII), it is proposed to examine in more detail some of the relationships in the organism-environment system that bear most directly on the main themes of tempo and mode.

THE NATURE OF ADAPTATION

It is a truism that all organisms can live under the conditions under which they do live and that they could not live under other sets of conditions that exist. To this degree, at least, and without any teleological implications, adaptation is universal. One group is better adapted to a certain environment than another if it is more successful in coping with the given conditions. From the viewpoint of evolution, the best criterion of such success is that the better-adapted group increases in numbers relative to the worse-adapted. Similarly, within groups a structural or physiological variant is a better adaptation if, on the average, individuals possessing it leave more offspring than those without it. This is not precisely what is usually understood by adaptation, but it is the essential pragmatic test of the evolutionary significance of adaptation. It is the basis for the theory of natural selection, but it could be a valid judgment of adaptation even if natural selection had, as some believe, no formative influence on the rise of new characters.

Survival and increase are not adaptation, and their use in measuring adaptation would be fallacious unless close correlation were demonstrated between them and adaptation. It will hardly be doubted that such a correlation does exist, but its degree in given cases may well be disputed. Some naturalists strongly deny that a hereditary character spreading in a population (because its possessors leave more offspring) is *ipso facto* adaptive, and they insist on a subjective criterion as to the usefulness of the character rather than on this more objective criterion which, they say with some justice, assumes what is to be proved. Indeed, the point has already been made on previous pages that inadaptive characters can theoretically spread and almost surely have in fact spread in populations under rather exceptional circumstances, but the evidence does indicate that the circumstances were exceptional.

The many inherent difficulties are illustrated by examples of characters with no apparent usefulness that are nevertheless associated with differential survival and hence may be suspected, at least, of being adaptive. For instance, there is abundant evidence that rather minor variations in scale characters of snakes, for which no usefulness has explicitly been demonstrated, are significantly associated with juvenile mortality (Dunn 1942). In other cases (possibly to some extent also in the example of snake scalation) it appears that the character studied is adaptive only in a rather oblique sense, not directly useful in itself, but the outward and visible manifestation of some more obscure character with far-reaching consequences for the organism (see, e.g., Haldane 1942, chap. ii). For instance, among young mice melanos have a higher death rate than their siblings of normal color. This fact was first demonstrated nearly a quarter of a century ago, but little notice seems to have been taken of it or of similar possibilities with broader implications. It is still widely assumed that because we study color visually, its adaptive function, if any, must be visual. Certainly color is sometimes a visual adaptation, but it does not follow that the factors of which color is one visible result are inadaptive or nonadaptive, as has so often been argued, when this relationship is not evident.

These and analogous examples show that adaptation is not a simple or single relationship and that the pragmatic test of the existence and relative value of adaptation does not adequately approach the problem of its real nature. Confusion on this point could be almost endlessly exemplified. Experimentalists are prone to use "adaptation" to mean modifications of phenotype induced by environment, or they may speak of genotypic "initial adaptation" and of induced phenotypic change as "adaptive modification." Phylogenists usually assume that adaptation is correlation of genetically determined functions with environment and that adaptability is the capacity for modifying the genotype in association with environmental changes. These and other usages reveal confusion of several theoretically and practically distinguishable factors.

It may be profitable to make three sets of distinctions in defining adaptation: between individual and group, between genotype and phenotype, and between the static condition of being and the dynamic condition of becoming. This leads to an eightfold classification of adaptation.

A. INDIVIDUAL
　I. Genotypic
　　1. Static: hereditary determination of possible organism-environment relationships
　　2. Dynamic: changes in this determination by individual mutation
　II. Phenotypic
　　1. Static: the existing relationship of organism and environment at a given time
　　2. Dynamic: ontogenetic changes in this relationship

B. GROUP
　I. Genotypic
　　1. Static: the distribution of genetic determinants of potential organism-environment relationships in a given population
　　2. Dynamic: the processes of change in this distribution
　II. Phenotypic
　　1. Static: the realized range of organism-environment relationships in a given population
　　2. Dynamic: the processes of change in this range

This classification and its practical significance for studying the phenomena of adaptation can be partially exemplified by application of Gause's experiments on adaptation and acclimatization of paramecia, considering a clone as a statistical individual (Gause 1942). Each clone was found to have a basic initial tolerance for salinity, A, II, 1. This tolerance was altered by acclimatization, A, II, 2. The genetic potentialities of the clone in this respect, A, I, 1, were thus approximately demonstrated. No change in these potentialities, A, I, 2, seems to have occurred during the experiment, so that there was no dynamic genotypic individual evolution.[1] The group of clones investigated showed considerable range in initial tolerance, B, II, 1, and in their changes under acclimatization, B, II, 2, demonstrating and approximately realizing population dispersion in pertinent genetic structure, B, I, 1. No new characters appeared, but in mixed populations some clonal genotypes spread at the expense of others, B, I, 2.

A reason for the difficulty and misunderstanding that can arise in attempting to relate the results of experiments like Gause's to the large-scale evolutionary phenomena studied by the paleontologist is the formulation of hypotheses and testing experiments simply in terms of "adaptation." Gause found that initial adaptation plus adaptive modifications was approximately constant among the various clones, but neither of the elements of this summation is quite what the paleontologist or field naturalist would label "initial adaptation" or "adaptive modification." Validity of analogy demands clarity of analysis into really homologous elements in the laboratory and in the field, and nowhere is the importance of definition more clearly shown.

REAL AND PROSPECTIVE FUNCTIONS

Such observations as those of Gause and such analysis of them emphasize the difference between the environmental relationship possible to an animal or group of given genotype and those actually realized. The distinction is clear to the point of being obvious, but it deserves repeated emphasis, especially for its bearing on phenomena like preadaptation. It is less familiar, at least in these words, but also important for outstanding modal phenomena of evolution that the same distinction may be made with respect to the environment: it, too, has relationships to organisms that are only potential and others that are

---

[1] This is clearly a crucial point in the problem of adaptation and adaptability and its absence was an unavoidable flaw in an otherwise very general and valuable series of experiments.

realized. Parr (1926) has clarified the subject by speaking of prospective and real functions of organism and of environment. The prospective functions of the organism are those that it could have in any environment. The prospective functions of the environment are those that are possible in it for any organisms. Real functions are those included in the common range or overlap of prospective functions of existing organisms and those of occupied environments (Fig. 24). The seemingly antagonistic principles of Darwinian selection and of the mutational preadaptation stressed by Parr and others are both involved in this situation. The environment determines what prospective functions of the organism will be realized, and the heredity of the organism determines what prospective functions of the environment will be realized. Here, again, distinction must be made between this condi-

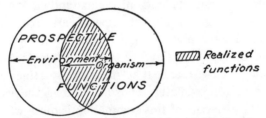

Fig. 24.—Diagram of the realization of functions by overlap of prospective functions of organism and environment.

tion, static from an evolutionary point of view, of the adaptive relationship of given organisms in given environments and the dynamics of change in adaptation in the course of phylogenetic history.

A group has not only prospective functions in being but also prospective changes of function. Different environments encountered and changes in environment condition changes in realized functions of the organism, and they also help to induce changes in the prospective functions of the organism. These changes are limited and may be spontaneously induced by changes in heredity. The neo-Darwinian and preadaptation principles are apparently opposite, but they are simultaneous, not antagonistic. The first predominates in some evolutionary sequences, and the second in others. Disagreement in theory arises mainly from emphasis on one to the exclusion of the other and from overgeneralization on this basis.

A clue to the process is that in determining the realized functions of the organism in a static situation, the environment also acts selectively to determine which phylogenetically prospective functions may become presently prospective and capable of realization in future generations. Important generalizations of descriptive phylogeny can be usefully expressed in these terms.

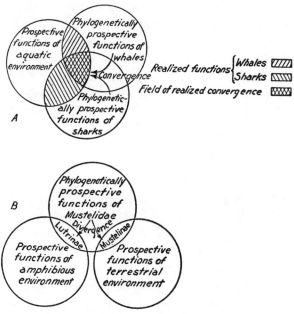

FIG. 25.—Convergence and divergence in evolution of phylogenetically prospective functions of organisms through overlap with the prospective functions of their environments. *A*, Convergence in evolution through overlap of some of the same prospective functions of an environment by phylogenetically prospective functions of two different groups of organisms; *B*, divergence in evolution through overlap of prospective functions of two different environments by the phylogenetically prospective functions of a single group of organisms.

Sharks and whales realize some identical or closely similar (and some quite dissimilar) functions of the marine pelagic environment. The primitively aquatic ancestors of the sharks could and did develop in divergent ways, among which one led to a realization of particular whale-like functions, just as among the primitively terrestrial whale ancestry one line of change in prospective functions led to the realization of shark-like habitus. The triple overlap of functions in the two types of organisms and the one type of environment has an area common to all three, which is the zone of convergence in sharks and whales (Fig. 25A).

The realization of divergent functions arises from the overlap of the phylogenetically prospective functions of one group of organisms on the prospective functions of two different environments. Examples are innumerable; for example, the fission of the Mustelidae into Lutrinae and Mustelinae is typical and clear (Fig. 25B). The ancestral mustelids did not have lutrine prospective functions in the static sense of Parr, but they had these in the dynamic sense that they were capable of developing them in the course of subsequent phylogeny. Development of such functions in two of the environmental conditions available to the group, amphibious and terrestrial, was the basis of divergence of the two subfamilies.

PREADAPTATION AND POSTADAPTATION

To enter a new environment, or, in Parr's terminology, to realize new functions, an organism must already have characters making it viable in that environment or corresponding prospective functions. This is preadaptation: the existence of a prospective function prior to its realization. Extremists of this school generalize (1) that preadaptation is random and independent of environment except a posteriori, and (2) that no function can be realized until after it is prospective and that therefore all adaptation is preadaptation. It may be taken as reasonably established that random preadaptation exists and has a definite role in evolution, but it is now equally well established that neither of these generalizations is universally valid.

Preadaptation is not necessarily, probably not usually, random or independent of environment. The direct development of adaptations in one environment may be preadaptive for another. This is repeatedly attested by change of function in animal history. The evolution of the crossopterygian fin was clearly adaptive in the aquatic environment. Its increasing mechanical strength and efficiency cannot reasonably be called random, independent of the environment, or preadaptive. But it reached a point where in addition to being an excellently adaptive structure for locomotion of a body suspended in water, it was also capable of supporting and pushing the body on a hard substratum. Then it became preadaptive with respect to a new environment. Its further evolution there was again adaptive, or postadaptive with respect to the crucial transition from swimming adaptation to walking adaptation. It was not at first a good walking mechanism, but only made this function barely possible. Millions of years of postadaptation were necessary before such a nearly perfect means of terrestrial locomotion as the horse limb was evolved.

Certain dental characters of the present Weddell's seal, *Leptonychotes weddelli,* can be theoretically interpreted as showing preadaptation and postadaptation on a more intimate and perhaps therefore more convincing scale (data from Bertram 1940). In allied seals, which do not normally use these teeth for cutting ice, the upper lateral incisors are somewhat enlarged and canine-like. In *Leptonychotes* this trend is carried noticeably farther, and these teeth are used, with the canines, to cut or saw breathing holes in the ice. These seals are thus able to pass the Antarctic winter beyond latitudes where there are many natural breaks in the ice, and so they survive in an environment not or less successfully open to any other mammal. (*Leptonychotes* ranges farther south than any other member of the Mammalia). Presumably this habit was not a factor in the evolution of these teeth to the degree found in allies of *Leptonychotes,* and this stage, doubtless ancestral for that genus, may be considered preadaptive for the function of ice cutting (although probably directly adaptive for other functions). This condition attained, the invasion of the new environment was made possible and the further evolution of these teeth in *Leptonychotes* may be considered postadaptive, developed by selection according to the requirements of the new habit. The intensity of selection may be judged by Bertram's competent opinion that the wearing out and loss of these teeth is the most important cause of death in the species. The frequency of damage to these teeth suggests that the adaptation is still incomplete and that further enlargement and greater durability may be expected to ensue rapidly (in a geological sense) if the genetic materials are available.[2]

The field naturalist who is familiar with the great conservatism of

[2] Which is probable, because the teeth show great variation, from near the minimum condition for viability to breeding age, toward the optimum, the optimum itself not yet being reached. Another factor does suggest, however, that the maximum selection pressure has been passed and that it will steadily decrease as the optimum is approached. Tooth failure and consequent drowning under the winter ice affects only the older individuals. As long as the animals affected are still able to breed, there will be some selection, but when the teeth become sufficiently durable so that they fail only in senile animals, selection will cease as regards this factor.

animal habits and psychology is not likely to be satisfied in such cases with the preadaptional axiom that animals enter a new environment simply because they can. It is probable that there is always some definite survival value involved in the change, and selection is thus also a decided factor in utilization of the preadaptive structure. In the nature of things, identification of the selective reason for the change is usually conjectural, but not necessarily pure speculation. In the case of the transition from crossopterygian to tetrapod limb, it is not believed that fishes began to walk because they could, but because those that could walk had a better chance of reaching water in which to live and breed when streams were drying up (Romer 1941, p. 47). Thus, among fishes living in these conditions, those best able to walk would leave most descendants: typical natural selection. In the case of Weddell's seal, wintering under the ice apparently makes this species less subject to attack by killer whales, which are a main cause of mortality among other seals but which cannot stay long beneath ice. So in almost all cases of change of environment, a reason for selective advantage can usually be found that is reasonable, even if not demonstrable.

There is a strong subjective element in insistence on universal preadaptation, because a character is not called preadaptive unless it becomes adaptive, which is usually, if not always, under the influence of selection and hence is not really preadaptation if the two are defined as logical alternatives. The difficulty disappears when it is recognized that preadaptation and postadaptation are phases of a single process in which selection is a conditioning factor throughout.

ADAPTIVE ZONES

The circumstances involved in survival and correlative with the characters of an organism in survival are exceedingly complex, and if these circumstances are called the environment, then environment is to be understood in the broadest possible way. It includes not only the physical conditions, average and variant, of the organism's geographic surroundings but also all existing foods, competitors and enemies, all forms of life affecting the given organism in any way whatever, other members of the same group, and even the organism itself, considered as an element in the total situation in which it exists. Clearly, no two animals ever have precisely the same environment, and no one environment ever remains the same from one instant to another.

Environments as a whole are divisible into an infinite number of classes by a large (but not infinite) number of criteria. Practical study of changes in adaptation, the most essential single phenomenon of modes of evolution, requires consideration of the environment as composed of a finite and more or less clearly delimited set of zones or areas. Such delimitation must have something of the arbitrary about it, and the sets chosen are sure to be different according to the problem studied or the individual students attacking them. To some extent, then, the subdivisions of environment represent a subjective pattern imposed on the real but unmanageable facts in order to simplify the latter sufficiently for comprehension by human minds in their present stage of evolution. In other words, classification and pattern in this field (to mention no other fields) are partly idealistic. Consideration of the philosophical implications of this admission would lead too far afield, but the admission needs to be made.

In principle there is no difficulty in accepting that the partly idealistic zoning of environments and of adaptive types has a realistic basis or equivalent. No one will doubt that sibling birds living in the same tree are, for all practical purposes, in exactly the same adaptive zone or that a fish in the sea and a bird in the air are in radically different adaptive zones in very fact and with no reservations as to idealistic simplification of pattern.

Races differentiated at, let us say, different but contiguous altitude ranges are zonally distinct only in the narrowest sense. A little more broadly, they show differentiation within a single adaptive zone. The statement is most strictly and literally true if, as may be, their ancestors occupied the whole zone without racial distinction. If, however, as may also be, one race developed from the other and concomitantly changed its range, one may still speak of differentiation within or more exactly of expanding occupation of one zone in this slightly broader sense. The adaptive shift is intrazonal or subzonal on this scale. On the other hand, a change from mainly aerial to mainly aquatic life, as in the evolution of the penguins,[3] is a major shift between totally distinct, essentially discontinuous adaptive zones on any scale.

Such discontinuity may seem an artifact in given cases, when only the environments are considered. From the point of view of both environments and organisms, of adaptation, the discontinuity is suffi-

---

[3] *Pace* a small minority of students who think that penguins' ancestors never flew.

ciently established as real when animals of intermediate habits are lacking. Canids and felids occupy different, discontinuous zones. Even the cheetah, a dog-like animal as cats go, does not contradict the fact that fully intermediate adaptive types would be anomalous in the established ecological system. I am fully confident that animals of completely intermediate structure between primitive true felids and primitive true canids did exist, but the fact that they did not persist and were evidently rare and local (no remains are known) is in itself evidence that there is a natural equivalent of discontinuity between the felid and the canid adaptive zones.

Each of these zones is a unit on the level of previous discussion. Within each there are some discontinuities of a lower order, for instance between machairodontines (the extinct sabertooths) and felines (all living cats and their extinct allies) among felids. For the most part, however, adaptive types within the well-defined and relatively large family zones are transitional, and feline evolution, for instance, can be pictured rather as deployment within a zone or a series of small subzonal differentiations than as involving transitions from one zone to another.

Above this, the terrestrial carnivores (fissipedes) occupy one larger adaptive zone, complex because it includes numerous such discontinuous lesser zones as those of canids and felids, and sharply discontinuous on its own level from the aquatic carnivore (pinniped) adaptive zone. Still higher in the complex is a carnivore zone distinguished, for instance, from a herbivore zone.

Similarly, the whole animal kingdom can be viewed as occupying a complex of larger and smaller adaptive zones, each definable in joint terms of the environments and of the organisms in them. The convenience of designating these in formal taxonomic terms ("canid zone," "felid zone," and so forth) arises from the approximation of adaptive, structural, and phylogenetic phenomena. Because of the widespread occurrence of convergence, this is only a convenience, not an equivalence. For instance, the thylacines, far removed from the Canidae taxonomically, belong in most respects in the same adaptive zone as the canids.

Adaptive zones, not only the animals occupying them, evolve. They evolve from physical changes, such as the wearing down of mountains or the spread of deserts in dry climatic periods. They also evolve because animals are part of the conditioning environment, and these animals evolve. For instance, there was no machairodontic adaptive zone until the evolution of moderate to large thick-skinned herbivores gave the sabertooth specialization decisive survival value. Then, after the zone arose, it changed in nature and requirements as the pertinent herbivores evolved. Finally, the extinction of many of the larger herbivores put an end to the machairodontine zone, which does not exist today, although it would probably persist, and we would probably have living sabertooths, if their zone had remained as it was in the Oligocene.

THE ADAPTIVE GRID

In accordance with these concepts, the course of adaptive history may be pictorialized as a mobile series of ecological zones with time

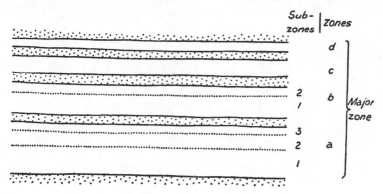

Fig. 26.—Diagram suggesting the complexity and nature of adaptive grid. The real grid is incomparably more complex than the diagram because it is not limited in number of dimensions or number and grades of subdivision.

as one dimension. Within the limits of the flat page, the basic picture resembles a grid, with its major bands made up of discrete smaller bands and these ultimately divided into a multitude of contiguous tracts (Fig. 26). It is understood that such a diagram is consciously oversimplified—for instance, the possibility of direct zonal shifts is not necessarily limited in nature to two, as in a printed diagram.

In the order in which such a grid system is normally occupied in the history of an actual group of animals, the lower bands or those first occupied tend to be wider. The zone of tolerance of fluctuating conditions is broader; adaptation is less specific; adaptability, in a phylogenetic sense, is greater. Successively higher bands, on the same scale, tend to be narrower until they may become so narrow, demanding such exact and specific adaptation, that a varying population cannot or can only temporarily fit them. They then correspond with a degree of specialization that is quickly lethal except under the most unusually static conditions. The whole pattern of nature is so varied that this and other features of the adaptive grid can be considered only tendencies, and all have exceptions. For instance, among primates man occupies

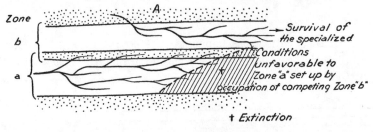

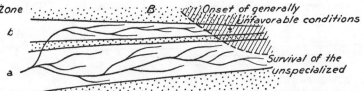

Fig. 27.—Diagrammatic representation on the adaptive grid of conditions leading to survival: A, The specialized; B, the unspecialized.

a relatively high zone, but it is broader than are those of many of the ecologically lower and earlier forms.

Forms within a single zone and to less extent those within separate contiguous zones compete if they are in geographic contact. Given a high degree of such competition under fairly stable extrinsic conditions, the normal result of such a situation is the replacement of a lower group by a higher (Fig. 27A); for example, the replacement of litopterns and notoungulates by artiodactyls and perissodactyls in South America. This is the survival of the specialized. Under conditions of rapid environmental change, however, or of intensified ecological stress, the capacity for survival tends to be proportional to the width of the adaptive zones. Occupants of the highest zones will become extinct first, those of the lowest zones last, if at all (Fig. 27B);

for example, differential survival in the Caenolestoidea. This is the survival of the unspecialized, previously discussed from another point of view.

The occupation of an adaptive zone depends on three factors. First, the zone must exist; it must represent a possible mode of life as regards environmental conditions extrinsic to the animals potentially or actively occupying it. This factor might be called preadaptation of the environment. Second, it must be unoccupied or occupied by a relatively less well-adapted group unable to defend it completely against new-

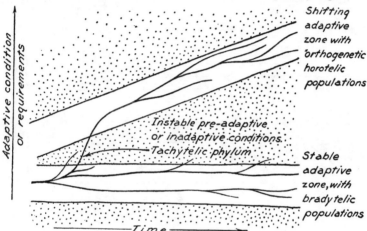

Fig. 28.—Diagrammatic representation on the adaptive grid of a bradytelic group, a tachytelic line arising from it, and the subsequent deployment and further evolution of this line as a horotelic group.

comers. Third, the group destined to occupy it must approach it so nearly that some variants become viable in it: preadaptation of the organism.

As the earth has become peopled, many new zones have become extrinsically available. For instance, an aerial insectivore zone, occupied variously by some pterodactyls, birds, and bats, was not present primordially, but arose only after the land had been occupied by relatively high types of plants and insects. Despite the ubiquity of life and the variety of organisms, there is no evidence that all possible, extrinsically existing zones are continuously occupied once they have appeared, although such lack of occupation may be exceptional. As far as the record shows, pterodactyls did not cede directly to their closest analogues, the bats, but became extinct long before bats arose and left this adaptive zone empty, unless it was marginally occupied by birds—a moot question. The mere fact that a group like the pterodactyls did become extinct suggests, but does not prove, that their adaptive zone became unavailable or ceased to exist for them. In this instance, climatic changes may have made added restrictions that were not present when the pterodactyls arose and that they did not meet. The zone then became available for bats, but the bats (or their preadapted insectivore ancestors) were not immediately available for it.

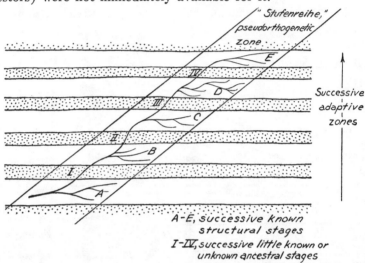

Fig. 29.—Diagrammatic representation on the adaptive grid of the step-like evolution of a group through successive occupation of different adaptive zones. The series *A-E* are *Stufenreihe* and may be taken as an orthogenetic series, although in fact the direction of evolution in each stage is not toward the next stage. This is, however, a reflection and to some extent an approximation of the undiscovered truly ancestral sequence, I-IV, the evolution of which is approximately rectilinear.

Rectilinearity in evolution, so-called "orthogenesis," is readily and, I believe, truly pictured as modal progression in phyla occupying a well-defined adaptive zone that is itself changing in time, the typical situation of horotelic evolution. In a similarly well-defined, but unchanging zone, the occupants, once well adapted, will tend to become bradytelic. The change from one zone to another is usually tachytelic (Fig. 28).

The pattern of step-like evolution ("Stufenreihe," Abel 1929 and elsewhere), appearance of successive structural steps, rather than directly sequent phyletic transitions, is a peculiarity of paleontological data more nearly universal than true rectilinearity and often mistaken for the latter. Examples could be drawn from every branch of paleontology, among them the European equids and the lung fishes (Dollo 1896). This important pattern at once suggests that of the adaptive grid (Fig. 29), and this similarity almost surely corresponds with a real relationship in nature. The steps preserved in the record represent the relatively abundant, relatively static populations of successively occupied adaptive zones, while the rare, rapidly changing transitional populations are, as a rule, absent from the record.

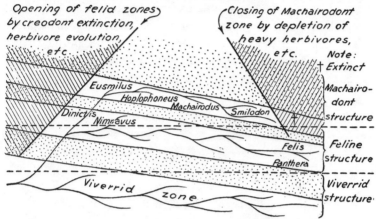

Fig. 30.—Greatly simplified representation on the adaptive grid of the evolutionary history of the Felidae. For fuller explanation see text.

By way of summary, and in part of extension, this grid method of pictorial analysis of evolution in an organism-environment system may be applied to the broader features of a real phylogeny, that of the Felidae[4] (Fig. 30). Their adaptive zone, as a whole, became available with the rise of various herbivores and the decline of earlier creodont carnivores. The felids, both feline and machairodontine, appeared rather suddenly, evidently derived from primitive, preadaptive viverrids. Incidentally, it is a characteristic part of the total pattern that

[4] Without worrying over details of exact generic lines of descent, which have no important bearing on these broader trends, I have followed essentially Matthew's conception (1910) of felid phylogeny, sustained by independent review of much of the evidence. It must, however, be mentioned that the equally authoritative opinion of Scott (e.g., 1937) upholds a radically different phylogeny, mainly on the grounds that Matthew's phylogeny is not strictly orthogenetic and involves some (merely nominal, as it seems to me) reversibility. Scott's phylogeny could be analyzed in terms of the adaptive grid with equal clarity, perhaps even with greater simplicity, but the picture would be different.

the lower viverrid zone persisted and is fully occupied today; within their geographic range the living viverrids are still more abundant and varied than the surviving felines, and they have outlived the highly specialized machairodontines. It happens that the shifting felid zone has been regressive in some respects: the early felines were in some (not all) respects more specialized in the machairodontine direction than later felines, and some early machairodontines (*Eusmilus*) represented a higher structural stage of that special type than did later forms. The effective machairodontine zone ceased to exist in the Pleistocene, and these animals became extinct, because of the extinction or wide depletion of the prey for which machairodontines could most successfully compete with other carnivores. Return to a more nearly feline adaptive zone, theoretically possible, was blocked by the full occupation of such a zone by the felines themselves.

Chapter VII: Modes of Evolution

THE SAME GENERAL FORCES are operative throughout the whole of evolution, and they bring about similar processes and sequences wherever and whenever they occur. Their predominance, balance, and interaction do, nevertheless, vary greatly, and quite different sorts of evolutionary patterns may result. These patterns are protean. Their seemingly infinite variety is so bewildering that generalization appears impossible at first. Yet through them all there run three major styles, the basic modes of evolution. Thus, despite their complexity and peculiarity in each case, almost all evolutionary events can be considered either as exemplifying one or another of these three modes or, more often, as susceptible to analysis as compounds of two or of all three.

Analysis destroys the compound. To say, as will be said, that an evolutionary pattern is composed of speciational, phyletic, and quantum elements may not be an adequate description of the whole pattern any more than, for instance, the statement that it is composed of calcium, carbon, and oxygen describes what limestone is like. Yet in both cases it will be agreed that the analysis, if correct, has revealed something of fundamental importance about the compound. There is, also, a subjective element in analysis, and (at present) much more in biology than in chemistry. Limestone may be analyzed as $CaO + CO_2$ correctly and usefully, as well as Ca-C-3O. In analyzing evolutionary patterns, different approaches may result in recognizing different elements, and it is impossible to say that any particular mode is elemental in any close analogy to the chemical sense of "element" or even that elements in this sense exist in the evolutionary compound.

The three evolutionary modes of this chapter do not, then, represent the only or the ultimate elements of evolutionary patterns. It is enough if they are established as one valid classification of basic descriptive evolutionary phenomena. They must be described as distinct things, separately typical or predominant at particular levels or in particular sequences; but it is to be remembered that any sequence at any level is almost sure to manifest a combination of modes and that a single mode perhaps does not occur in nature in wholly pure form.

The three modes that are to be distinguished, largely on the basis of all the discussion of tempo and of other evolutionary factors on previous pages, may be labeled speciation, phyletic evolution, and quantum evolution. This terminology leaves much to be desired, but no other is available and this will serve if it is taken as here defined and in its present context. Speciation, in this sense, is so called because it is a mode most clearly evident in the process of racial and specific

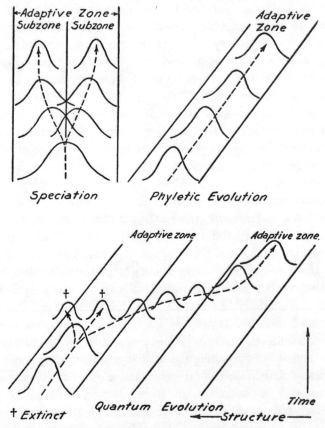

Fig. 31.—Diagrams of characteristic examples of the three major modes of evolution. In this and Figs. 32-33 the broken lines represent phylogeny and the frequency curves represent the populations in successive stages.

evolution, but again it is emphasized that the mode is not confined to or the only one at these levels. Phyletic evolution is here so called because it is dominant in the pattern of more or less linear, or better "path-like," evolution of most phyla[1] as they are seen by the paleon-

---

[1] "Phylum" in this sense, a direct sequence of ancestry and descent at any taxonomic level, is different from "phylum" in the purely taxonomic sense of a major division of

tologist. Quantum evolution, which like the other modes occurs at all levels, but is most important and distinctive at relatively high levels, is so called because it typically results from a sort of all-or-none reaction; an evolutionary change, definite in quantity, is either completed at a relatively high rate or it is not completed at all, and the population involved simply dies out.[2] Generalized examples of the three modes are shown in Fig. 31, and each will now be discussed in more detail.

SPECIATION

The process typical of this mode of evolution is the local differentiation of two or more groups within a more widespread population. On the smallest scale, this process involves group differences so minor and so fluctuating that they are given no taxonomic designation and have no clear evolutionary significance, although the event may prove that they are the beginning of changes that do become permanent and important. At a slightly higher level local groups attain temporary equilibrium, but the condition is not fixed or irreversible; the groups are approximately subspecies in taxonomic terms. If, or when, definite isolation of the groups occurs, there is a splitting of the population into two or more separate closed systems, which are at first species, but may by a continuation of the same sort of process and other processes become genera or somewhat higher units. The evolutionary significance of the distinction between genetically nonisolated and isolated population units is great and has been sufficiently stressed. From the point of view of pattern, the distinction also exists; but the two may well be considered phases or varieties of one general mode. The essential similarity of pattern is particularly clear in paleontological and zoological studies on the basis of morphology, without available genetic criteria, in which the pattern difference is one more of degree than of kind. In such studies the word "speciation" generally is meant to include the process of subspecies formation, "raciation," as well as that of species formation, strictly speaking.

the animal kingdom. Both usages are current and proper and they should hardly be confused in their contexts, but such confusion has occurred and has incited some polemics on evolutionary theories.

[2] Apologies may be in order for borrowing a term now so popular in physics and using it in a sense only distantly and imperfectly analogous with that of a physical quantum. The term is much older than "quantum physics," however, and its general meaning is as applicable to this evolutionary mode as it is to the quantum of Planck.

In relation to the adaptive grid, this process occurs within one zone and is subzonal. Two intergrading and often superimposed sorts of detailed processes may be distinguished in this respect (Fig. 32). One may be visualized as the differentiation of a population spread over a zone into subzonal units, and the other as the deployment or fanning out of a population across adjacent subzones of one zone. In both cases the adaptive factor is adjustment to relatively minor differences in local ecological conditions. In the first instance, adaptation becomes narrower and more particularly for a given niche. In the second in-

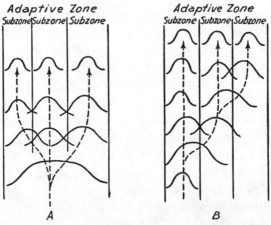

Fig. 32.—Two patterns of speciation: *A*, a single widespread population becoming differentiated into more specialized, locally adapted groups; *B*, a local population spreading into adjacent subzones, to each of which descendents become specially adapted.

stance, there are local changes of exact adaptive type. In addition to the strictly adaptive changes, random segregation may occur in both; but within this mode it is seldom a dominant factor.

The direction of evolution in speciation as a mode (rather than strictly as the rise of a species) is shifting, erratic, and not typically linear. The simple local differentiation of a widespread population can hardly be said to have direction; it may be a concentration rather than a directional movement. Although reversion does not necessarily or usually occur, it remains possible. Differences between nonisolated populations may be obliterated by genetic interchange. Even though groups have become genetically isolated, they long remain genetically similar, both as regards the prevailing genotypes and as regards potentialities for change, and this also leaves their evolutionary distinctions essentially reversible.

The resulting pattern of actual streams of genetic descent has little resemblance either to the classic tree of life or to the linearity of parallel or radiating phyla. It is rather a complex, almost amorphous, network formed by anastomosis in the finer mesh and multiple fission in the coarser pattern. Long-range stability is perceptible, if at all, only in the broadest relation to the whole population. In detail, in the sequences really illustrating the mode of speciation there are only temporary equilibria, repeatedly changing, with greatest flexibility in minor adjustments.

This sort of differentiation draws mainly on the store of pre-existing variability in the population. The group variability is parceled out among subgroups, or a lesser group, pinched off from the main mass, carries with it only part of the general store of variability. The parceling-out phase commonly results in a reduction of variability available within any one local population. This is frequently followed by an expanding phase of one or more of the most successful local populations, during which little evolution in the sense of change in average characters occurs, but the depleted variability is replenished and may again be expended in reserved differentiation. Differentiation may be assisted by, but does not require, the incidence of new mutations during the process itself.

Speciation, in this sense, is more likely to be a matter of changing proportions of alleles than of absolute genetic distinctions, although the latter also occur. It follows, as is abundantly confirmed by morphological taxonomy, that the phenotypic differences involved in this mode of evolution are likely to be of a minor sort or degree. They are mostly shifting averages of color patterns and scale counts, small changes in sizes and proportions, and analogous modifications.

The populations involved may be of any size, but the most common type, or the one most propitious for this mode of evolution, is apparently of moderate size. The size of the pertinent units is likely to diminish during the actual phase of differentiation, although the population as a whole need not do so and may even increase. Isolation may be absent and is not necessary for or typical of this mode, although commonly occurring within it.

As far as such evolution depends upon segregation of existing variations, it may be very rapid or, as an unusual limiting case, practically instantaneous. It cannot, however, persist long or carry far at a high rate. As a sustained phenomenon, this mode implies moderate average rates over long periods, probably comparable to the horotelic distributions of phyletic rates. In detail, individual rates appear to be quite erratic.

If these processes are long continued, the mode changes and grades into that of phyletic evolution, which is, nevertheless, distinguishably different in its typical expressions, as will be discussed. The accumulation of the changes of the speciational mode tends toward segregations of higher category, but such accumulation also tends to be of a different sort. The accumulation of adaptive characters leads to sequences in the second mode. The accumulation of nonadaptive characters tends either to extinction or, under special circumstances, to sequences in the third mode.

This first category is almost the only mode accessible for study by experimental biology, neozoology, and genetics. It embraces almost all the dynamic evolutionary phenomena subject to direct experimental attack. It is also subject to paleontological study, but at present such work is only beginning and the extent of its possible contribution is not determined.

A few examples of speciation, strictly speaking, have been noted in previous chapters. Innumerable others are given in almost all works on the principles of evolution in general and in a host of taxonomic monographs, some of which are known to all students of the subject and require no summary here.

PHYLETIC EVOLUTION

The evolutionary mode to be discussed as phyletic involves the sustained, directional (but not necessarily rectilinear) shift of the average characters of populations. It is not primarily the splitting up of a population, but the change of the population as a whole. Obviously it can give rise to new species just as well as the different mode here called speciation, but it is not the typical mode of speciation, and it is less clearly seen at that level than at higher levels. The phyletic lines that fall into patterns of this mode are composed of successive species, but successive species are quite different things from the contemporaneous species that are involved in speciation as that word is

Modes of Evolution

usually used. The lines themselves and a fortiori the groups of allied lines among which phyletic evolution is usually observed are of greater than specific value, and this mode is typically related to middle taxonomic levels, usually genera, subfamilies, and families.

In relation to the adaptive grid, phyletic evolution is usually or most clearly seen as progression of single or multiple lines within the confines of one rather broad zone. It is basically zonal, as opposed to the subzonal character of the speciational mode. Subzonal speciation, of course, proceeds continually within the zone simultaneously with the zonal phyletic evolution.

Aside from isolated discoveries that contribute less directly to the study of evolution, nine-tenths of the pertinent data of paleontology fall into patterns of the phyletic mode. It has naturally resulted that paleontologists have overemphasized this mode and have overgeneralized on the basis of it, just as most experimentalists have overemphasized and overgeneralized from the speciational mode. Nevertheless, the phyletic mode is one of very wide occurrence, consequently of major importance, and the abundance of paleontological data makes it one of the best known.

It is within this mode that evolution tends to be most strictly adaptive. Hence the overgeneralization, e.g., of Osborn, that all evolution is adaptive. There is little or no random change. Adaptively unstable marginal variants do, indeed, appear but in the usual pattern they do not persist, presumably being either eliminated by selection or diluted or lost as phenotypes by back breeding. Changes are usually either by postadaptation, the process of specialization by perfecting adjustment to an occupied ecological position, or by secular adaptation, relatively slow adjustment to a shifting environment or change from one environment to another, generally similar, by stages in which adaptive equilibrium is constantly maintained.

It is inherent in these processes that they commonly appear directional. The existence of a trend is usually clearly discernible, and again this repeated observation has been overgeneralized in the dictum that all evolution has definable trend. This trend is not necessarily rectilinear. In fact, as previously exemplified, it seems to be very rarely truly rectilinear in detail and for long periods of time. It does, however, often appear as approximately rectilinear when viewed more broadly as the direction of average change in numerous different characters of various ecologically or phylogenetically allied groups. The possibility of reversion is sharply restricted, and as a matter of observation reversion in this mode is unusual and probably never complete. (The source of another overgeneralization in the "law" of irreversibility of evolution in general.) This tends to increase the linearity of patterns and to make them approximately rectilinear, at least for more restricted periods of time. Anastomosis of phyletic lines above the levels of pure speciation is absent. Branching occurs and is even common, but cannot be called typical. Branching is usually by bifurcation; when

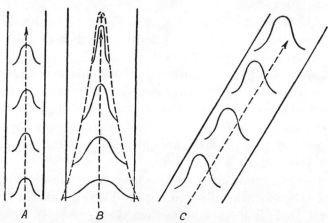

FIG. 33.—Three patterns of phyletic evolution: *A*, a population early well-adapted to a stable zone with little subsequent change; *B*, a population becoming narrower in adaptation, more highly specialized, and eventually extinct; *C*, a population slowly changing in response to a shifting adaptive zone.

it is multiple, the number of branches is less than is characteristic of many examples of speciation.

Within this varying mode the following lesser sorts of patterns stand out as particularly clear and important.

(1) Early adequate (postadaptive) response to relatively stable environment, followed by an indefinite period in which selection pressure is centripetal and evolutionary changes are small (Fig. 33A).

(2) Increasing linear intensification of adaptation, building up to a rising adaptive peak, with higher and higher specialization for more and more restricted environmental conditions, finally terminated by extinction when those conditions change more rapidly than the specialized group can follow (Fig. 33B).

(3) Shift in adaptive type, particularly in response to a changing environment (Fig. 33C), also at times by deployment into lesser ecological stations (analogous, but not identical with speciation and essentially a multiple of *b*), or by invasion of marginal environments. This last subtype of phyletic evolution intergrades with the next mode, quantum evolution.

In all three subtypes, and in phyletic evolution generally, the organism-environment complex is essentially stable. If for any reason instability appears, it is either almost simultaneously compensated for by population response, the normal situation in sustained phyletic evolution, or it changes the mode, which becomes quantum evolution or leads to extinction.

Variability is maintained at a nearly constant level. Rates of evolution are low enough to allow constant replacement of variation that might otherwise be depleted in genotypic progression, depletion by multiple segregation is not here an important factor, and radically new variants and excessive variation are held in check by selection, which is a dominant factor within this mode.

The sorts of phenotypic characters involved in phyletic evolution are, in general, the same as in speciation. They are usually characters of size, proportions, and number or differentiation of isomeric elements (polyisomerism and anisomerism of Gregory). The difference from speciation in this respect is mainly that phyletic evolution, as a process rather secular than periodic, finally produces changes considerably greater in degree. Radically new morphological types, markedly remodeled reaction systems, are rarely produced, and, if they ever are, only over extremely long intervals of geological time.

Phyletic evolution goes on continuously, along with speciation, in populations of all sizes and kinds, but it is most typical of and most clearly seen in groups with continuously moderate to large breeding populations. Commonly it is a pattern spread over numerous separate, related lines, deployed geographically and ecologically within a broad adaptive zone. Each such line is usually completely isolated in a genetic sense, but within each the local population is often large and there is normally strong genetic migration.

It is primarily from data for phyletic evolution that the existence of bradytelic and horotelic rate distributions was inferred, and both are characteristic of phyletic evolution. As previously noted, bradytelic rates are less common than horotelic. They arise within the subtype of phyletic evolution designated as (1) above.

Numerous instances of phyletic evolution have been mentioned on previous pages, notably the stock example of the Equidae. Osborn's many monographs, although they may overgeneralize on this basis, are particularly valuable for their wealth of illustrative material on phyletic evolution. Practically all the paleontological studies already cited and many others give excellent examples and concrete data.

QUANTUM EVOLUTION

Perhaps the most important outcome of this investigation, but also the most controversial and hypothetical, is the attempted establishment of the existence and characteristics of quantum evolution. A "quantum," in a sense more general than but including the definition of physics, is a prescribed or sufficient quantity. The term is applicable in situations in which subthreshold actions produce no reactions but superthreshold actions produce reactions of definite (not necessarily equal) magnitude (this magnitude being strictly the quantum involved). For the sake of brevity, the term "quantum evolution" is here applied to the relatively rapid shift of a biotic population in disequilibrium to an equilibrium distinctly unlike an ancestral condition. Such a sequence can occur on a relatively small scale in any sort of population and in any part of the complex evolutionary process. It may be involved in either speciation or phyletic evolution, and it has been mentioned that certain patterns within those modes intergrade with quantum evolution.

It is the present thesis that quantum evolution also occurs on a larger scale and in clear distinction from any usual phase of speciation and phyletic evolution. Attention will here be focused on this more distinctive and more important sort of quantum evolution. Like the other modes, it can give rise to taxonomic groups of any size, and the sequences involved can be (subjectively) divided into morphological units of any desired scope, from subspecies up. It is, however, believed to be the dominant and most essential process in the origin of taxonomic units of relatively high rank, such as families, orders, and classes. It is believed to include circumstances that explain the mystery that hovers over the origins of such major groups.

The origin of such a group involves the rise of a distinctly new adaptive type or, on the adaptive grid, the shift from one main zone to another. In these terms, quantum evolution is interzonal, in contrast with zonal phyletic and subzonal speciational evolution. The essential difference, however, is not so much the transfer from one zone to another, which may also occur in phyletic evolution, at least as regards movement between approximated zones, but a form of discontinuity. In phyletic evolution equilibrium of the organism-environment system is continuous, or nearly so, although the point of equilibrium may and usually does shift. In quantum evolution equilibrium is lost, and a new equilibrium is reached. There is an interval between the two equilibria, the biological analogue of a quantum, in which the system is unstable and cannot long persist without either falling back to its previous state (rarely or never accomplished in fact), becoming extinct (the usual outcome), or shifting the whole distance to the new equilibrium (quantum evolution, strictly speaking).

Such an event can be divided into three phases: (1) an inadaptive phase, in which the group in question loses the equilibrium of its ancestors or collaterals, (2) a preadaptive phase, in which there is great selection pressure and the group moves toward a new equilibrium, and (3) an adaptive phase, in which the new equilibrium is reached. That it involves preadaptation as a necessary element is another marked distinction between quantum evolution and phyletic evolution, which is mostly adaptive (and postadaptive), and speciation, which is largely adaptive and in part random nonadaptive.

The evidence that such evolutionary events do occur has, for the most part, been summarized in earlier chapters. The bulk of this evidence is of three sorts: first, recorded paleontological sequences of quantum evolution of a relatively minor, transitional type; second, paleontological evidence, mostly indirect, but to my mind convincing, that major transitions do take place at relatively great rates over short periods of time and in special circumstances; and, third, experimental and observational evidence in the field of population genetics that such a mode of evolution has a probable mechanism and would be expected under given conditions. It has also been pointed out that the most disputable point is whether the transition is instantaneous, in closer analogy with the quantum of physics, or whether quantum evolution occurs at rapid but finite rates. Strong reasons have been adduced for believing that the latter is the case. It is unnecessary to repeat these considerations in the present chapter. It is, however, proposed to make the nature of quantum evolution clearer by an example before proceeding with a more general discussion of its typical circumstances and characteristics.

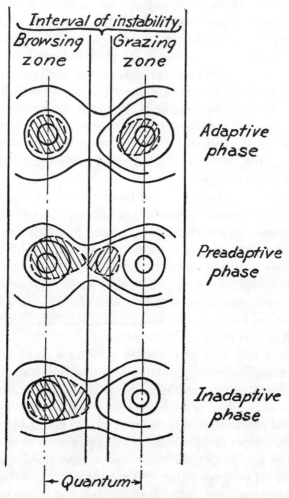

Fig. 34.—Phases of equid history interpreted as quantum evolution (see also Fig. 13). The phase designations refer to the part of the population breaking away and occupying the grazing zone.

The example chosen (Fig. 34 and the following text) does not illustrate quantum evolution in its purest and most important form, as a radical transformation of type leading to the rise of a new high group

Modes of Evolution

category of animals, but is a somewhat marginal case in which quantum evolution and phyletic evolution intergrade. The example is clear and is abundantly documented. For reasons discussed in Chapter III, major examples of quantum evolution are never well documented—indeed this deficiency of documentation is itself good evidence for the reality of the inferred phenomenon.

In discussing equid hypsodonty it was shown that there are two main adaptive zones, each embracing a shifting adaptive peak—one corresponding with browsing habits and one with grazing habits. In the earlier stages only the browsing peak or zone was occupied by the ancestral horses. The peak shifted (moved along the oblique zone in time) as the animals increased in size. This change in the organism-environment system made increase in tooth height adaptively valuable. The slow and steady increase was fully typical of phyletic evolution in its most nearly pure form. An effect of this change was that the browsing peak approached the as yet unoccupied grazing peak. There was considerable variation at all times, evidently with mutations both toward and away from hypsodonty. Selection pressure against hypsodonty above the optimum existed, but was relatively weak, because this condition was unnecessary rather than decidedly disadvantageous, such as hypsodonty much below the optimum. Some of the more hypsodont variants reached a point on the adaptive landscape at the base, or on the lowest slopes, of the grazing peak. It became possible for them to supplement their food supply by eating some grass, a relatively harsh food and highly abrasive to the teeth, although if they had eaten only grass, their teeth would not have outlasted the normal reproductive period, and they would have been opposed by very strong selective pressure.

This point was a threshold. It initiated a sort of trigger effect that set off an evolutionary quantum reaction, although this differed from a major quantum effect and partook also of the aspects of phyletic evolution, because it occurred within a continuously large population and involved only a relatively simple change in a few organs rather than a transformation of the whole organism.

That part of the population that had reached the slope of the grazing peak was no longer in the centripetal selection field that surrounded the browsing peak. On the contrary, it began to be affected by strong linear selection toward the grazing peak. The variation that brought them to this point was essentially inadaptive, because it was distinctly beyond the optimum of the peak occupied by the population to which they had, up to this time, belonged. They were now imperfectly adapted either for browsing or for grazing, but they were preadapted for grazing. The outcome was that this segment of the population, in the stages centering around *Merychippus*, evolved with relative rapidity into fully grazing forms. That is, they broke away from the population around the browsing peak (which continued to be occupied by such forms as *Anchitherium*) and moved up the grazing peak. Later forms, in the postadaptive phase, became closely concentrated on the summit of this rising peak.

On a relatively small scale, the distance from the browsing to the grazing peak in this example is a quantum, a step that must be made completely or not at all, although in the example and, I think, in general the genetic processes involved do not permit making the step with a single leap, and the selective processes do not make the unstable intermediate forms inviable.

The pattern involved in this and in other instances of quantum evolution is more rigidly linear than in any other mode of evolution and must tend to be almost strictly rectilinear, but of relatively short duration. In the normal case, at least, reversion is virtually impossible once a group has become unstable and has separated from the main body of the ancestral population. The group must then go on or become extinct unless very exceptional circumstances remove the disadvantageous features of the inadaptive condition, which, however, is almost equivalent to their ceasing to be truly inadaptive.

It is clear that populations in any stage of quantum evolution but the earliest and the latest may be decidedly unstable. The transition is a phase of nonequilibrium, and groups in this genetic and structural intermediate condition do not long survive unchanged. Although transitional stages of speciational evolution are abundantly represented among contemporaneous forms and those of phyletic evolution are not uncommon, transitional stages between groups that were differentiated by quantum evolution are extinct, apparently without exception.

A necessary condition for quantum evolution is the appearance of preadaptation, which in the typical case, although not in all cases, means the fixation of characters that are inadaptive or at most nonadaptive in the circumstances under which they first appear. As out-

Modes of Evolution

lined in Chapter II, the condition most propitious to such fixation is that the population be small and completely isolated. Mutation rates must be at least moderate, and extraordinarily high rates would promote the process, but are not necessary to it and probably not typical of it. Mutations with radical phenotypic effects, or unusually large mutations, would unquestionably provide a theoretically excellent mechanism for such transitions. As far as is known, however, such mutations are wholly random with respect to adaptation, and this makes the probability of their producing a viable preadaptation so extremely small that it is hardly conceivable that they have any important role in the normal processes of evolution. As has repeatedly been remarked, for instance by Dobzhansky (1942), also on previous pages of this study, the accumulation of small mutations is not only adequate to permit rapid evolution, such as is involved in quantum evolution, but also theoretically the best substantiated mechanism for this.

Although the most radical types of quantum evolution probably begin by random fixation of inadaptive mutations in very small, isolated populations, there are at least two other situations that can give rise to such sequences—usually on a lesser scale and with less clearcut distinction from the speciational and phyletic modes. The loss of adaptive equilibrium that initiates a quantum change need not arise, or not wholly, on the organism side of the organism-environment system but can also be the result of a change in the environment which, if more rapid than the population can follow on the basis of existing variability, also effectively upsets the equilibrium. This unfavorable situation must in itself tend to cause the reduction of the population as a whole or its fission into very small local groups, thus producing the typical conditions for quantum evolution. It is also possible, as explained and exemplified in Chapter VI, for preadaptation to arise without any positive inadaptation, so that equilibrium is nearly or quite maintained up to the point when selection pressure toward the new condition becomes effective. In such cases there is never any strong pressure against the new trend and the instability of the transitional populations is only relative. This phenomenon, similar to the example of horse hypsodonty treated above, is a frequent concomitant of phyletic evolution and grades imperceptibly into patterns that are purely phyletic, not properly examples of quantum evolution.

In all typical cases in which the transitional stages are unstable, the vast majority of populations that fall into such a condition do not, in fact, make a quantum step to new stability, but simply become extinct. This pruning process, a major, but of course not the only, aspect of natural selection, goes on constantly in every group. Among a thousand small local populations that may become unstable from fixation of inadaptive mutations or from change in environmental conditions, perhaps only one or perhaps none will actually proceed by quantum evolution. Besides the simple existence of such a group, there are two other necessary conditions, the coincidence of which is unusual, although it has occurred many times among the millions of biotic groups over the millions of years in geological history. Materials for further advance in the given direction must be provided in the form of repeated, progressive mutations. Regarding this there is little to say at present, except that it does occur. Doubtless it also frequently fails to occur and this puts an end to the group in question. Moreover, it is necessary that there should be in fact a new adaptive zone available and open to the evolving group. In other words, quantum evolution occurs only if there is preadaptation for an effectively possible new mode of life.

The availability of a new adaptive zone does not depend alone on its physical existence and proximity, but also on its being open to other occupants (that is, empty) or so sparsely or marginally occupied that it involves no great competition. This is the more true because the small, unstable populations typical of the transitional phase of quantum evolution are especially liable to extinction and especially disqualified for meeting any strong competition from populations already entrenched in a zone. In the history of life it is a striking fact that major changes in the taxonomic groups occupying various ecological positions do not, as a rule, result from direct competition of the groups concerned in each case and the survival of the fittest, as most students would assume a priori. On the contrary, the usual sequence is for one dominant group to die out, leaving the zone empty, before the other group becomes abundant and occupies any position in the zone except possibly the marginal fringes. Ichthyosaurs became extinct millions of years before their mammalian ecological analogues among the Cetacea appeared. Pterodactyls were long absent before bats occupied a similar or overlapping zone. Dinosaurs became extinct before the larger terrestrial mammals so quickly radiated into much the same spheres. In the latter example (but not the other two) the replacement followed so quickly after the disappearance of the older group that it is reasonably inferred that mammals, in the form of multiple, small, local exploratory groups, had, so to speak, been lapping against the base of the dinosaur adaptive peaks for some time when the otherwise irrele-

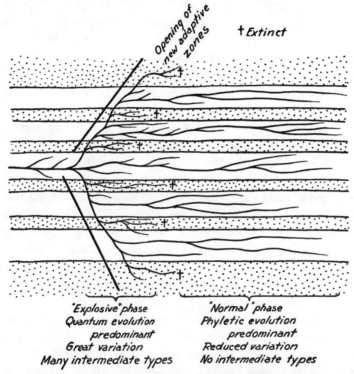

Fig. 35.—Diagram of "explosive" evolution by multiple quantum steps into varied adaptive zones, followed by extinction of unstable intermediate types and phyletic evolution in each zone. The pattern is like that of South American ungulates, although the diagram does not attempt to show their actual phylogeny in detail.

vant extinction of the dinosaurs suddenly permitted the mammals to climb those peaks. In such a case the trigger action that sets off the quantum reaction is something extraneous to the populations taking part in the reaction.

Such an opening of major zones at a time when there are existing groups to some extent preadapted for them is advanced as an explanation of the phenomenon of so-called "explosive" evolution, repeatedly observed and emphasized by paleontologists, but usually imputed to some sort of evolutionary exuberance, "youthfulness," excessive mutation, or the like, within the group concerned (see Fig. 35).

Among many examples of such an explosive phase of evolution, that of the native South American ungulates is unhackneyed and happens to be relatively familiar to me at firsthand. At the beginning of the Tertiary their ancestors were (by inference—they are not objectively known and are not likely to become so if the inference is correct) scattered, small local groups of obscure and diminutive animals. Invasion of a new domain, having many then-unoccupied adaptive zones for which they had incipient preadaptation, was followed by an enormous increase in their numbers and their rapid occupation of numerous zones by quantum steps; in some cases they jumped repeatedly from one zone to another that had not been available until the former had been occupied. The earliest well-known mammalian fauna in South America, the Casamayoran, presents the unusual and fascinating spectacle of such a great deployment still in its quantum phase, although nearing the end of it. Intragroup variation is enormous, intergroup variation is also unusually great in some cases, the ungulate population is protean almost beyond parallel, and there are so many transitional types, some of them quite plainly adaptively unstable, that the taxonomic problem seems almost hopeless. In successive later faunas all the transitional types rather quickly drop out, variation decreases, and the ungulate fauna settles down into a limited number, fewer than a dozen, of well-defined and well-separated major ecological zones. There was no subsequent quantum evolution, except in the smallest and least clear-cut way, and in each zone the rather stereotyped surviving lines followed the comparatively slow and staid course of phyletic evolution, complicated in detail by continually recurrent speciation, until they became extinct (as all did eventually). In sum, the whole history of ungulates in South America is a remarkably instructive example of the occurrence and interplay of all three major modes of evolution.

Although not exemplified by the South American ungulates, the phenomenon of so-called senile variability is also related to similar elements in the evolutionary pattern. The idea of racial old age as an analogue of individual old age is quite unacceptable to me and seems more a metaphysical notion than an expression of a physical relationship. Yet it does appear to be true that some sorts of animals, by no means all, showed definite increase in intergroup variation before becoming extinct. It seems probable that this phenomenon is a concomitant of the increasingly unfavorable environmental conditions that eventually caused the extinction. From this point of view, the increased

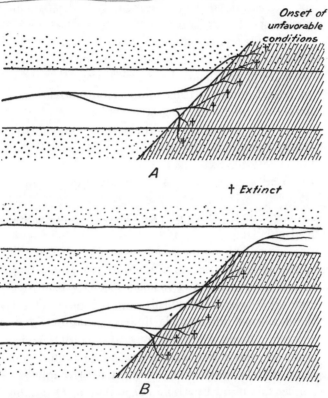

Fig. 36.—Intergroup variation induced by the onset of unfavorable conditions: *A*, condition in which there is no adjacent zone available and the whole group becomes extinct; *B*, condition in which such a zone is reached by the quantum evolution of one line, giving rise to a new major zoological type.

variation, even though it did not prevent extinction, may be considered an evidence of racial vitality rather than of senility, although "racial vitality" is also a vague term with metaphysical overtones. The unfavorable environmental changes produce population attenuation and subdivision with small local groups adaptively unstable and prone to fixation of inadaptive mutations. The situation as a whole is inherently

TABLE
CHARACTERISTICS OF THE MAIN

Mode	Typical Taxonomic Level	Relation to Adaptive Grid	Adaptive Type	Direction	Typical Pattern
Speciation	Low: sub-species, species, genera, etc.	Subzonal	Local adaptation and random segregation	Shifting, often essentially reversible	Multiple branching and anastomosis
Phyletic evolution	Middle: genera, subfamilies, families, etc.	Zonal	Postadaptation and secular adaptation; (little inadaptive or random change)	Commonly linear as a broad average, or following a long shifting path	Trend with long-range modal shifts among bundles of multiple isolated strands, often forked
Quantum evolution	High: families, sub-orders, orders, etc.	Interzonal	Preadaptation (often preceded by inadaptive change)	More rigidly linear, but relatively short in time	Sudden sharp shift from one position to another

and increasingly unstable and tends to a multiplicity of types of structure that might be called exploratory, without teleological intent. The situation is ideal for quantum evolution, but this may not occur except temporarily on a small scale, if the specialization of the group precludes the rise of adequately preadaptive new genetic structures, or adjacent, open adaptive zones are not present, or both. In such a case the phase of "senile" variation can be considered simply an explosive phase that failed for lack of places to explode into (Fig. 36A).

On the other hand, in what seems in other respects to be quite the same sort of situation it can happen that one or more of the exploratory populations does succeed in invading a new zone, and then a radically new organic type arises by quantum evolution (Fig. 36B). For instance, not long before the extinction of the mammal-like reptiles, at the end of the Triassic, they seem to have undergone increasing environmental stress and to have responded, in part, by fractionation of populations and increased intergroup variation. Among the

19
MODES OF EVOLUTION

Stability	Variability	Typical Morphological Changes	Typical Population Involved	Usual Rate Distribution
Series of temporary equilibria, with great flexibility in minor adjustments	May be temporarily depleted and periodically restored	Minor intensity; color, size, proportions, etc.	Usually moderate with imperfectly isolated subdivisions	Erratic or comparable to horotelic rates
Whole system shifting in essentially continuous equilibrium	Nearly constant in level; most new variants eliminated	Similar to speciation, but cumulatively greater in intensity; also polyisomerisms, anisomerisms, etc.	Typically large isolated units, with speciation proceeding simultaneously within units	Bradytelic and horotelic
Radical or relative instability with the system shifting toward an equilibrium not yet reached	May fluctuate greatly; new variants often rapidly fixed	Pronounced or radical changes in mechanical and physiological systems	Commonly small wholly isolated units	Tachytelic

probable thousands of such groups, there was one (or, probably, there were several) that made the quantum step to the mammalian adaptive zone. The others all became extinct, and with them the mammal-like reptile structure as such. An event like this, although a "senile" phase from the point of view of the antecedent groups, differs from a "juvenile" explosive phase only in that the new zones occupied are fewer in number and not so immediately conducive to great increase in numbers and in variety of types.

SUMMARY

By way of summary, certain of the conditions, factors, and characteristics believed to be most typical of the three main modes of evolution are tabulated. In each case it is not implied that the stated condition for a given mode is the only possibility within that mode or that it is confined to that mode, but only that it is a common, favorable, or distinctive condition.

Works Cited

Abel, O. 1928. Das biologische Trägheitsgesetz. Biol. Gen., 4: 1–102.
—— 1929. Paläobiologie und Stammesgeschichte. Jena, Gustav Fischer.
Bailey, J. L., Jr. 1941. A contribution to the theory of evolution by natural selection. Amer. Nat., 75: 213–230.
Basse, E. 1938. Sur une nouvelle espèce de *Valenciennesia* (pulmoné thalassophile) du Cénomanien malgache: *V. madagascariensis* n. sp. Bull. Mus. Hist. nat. (Paris), 10: 659–661.
Bertram, G. C. L. 1940. The biology of the Weddell and crabeater seals. Brit. Mus. (Nat. Hist.), Brit. Graham Land Exped., Sci. Repts., I, (No. 1): 1–139.
Blair, W. F. 1941. Techniques for study of mammal populations. Jour. Mammalogy. 22: 148–157.
Brinkmann, R. 1929. Statistisch-biostratigraphische Untersuchungen an mitteljurassischen Ammoniten über Artbegriff und Stammesentwicklung. Abh. Ges. Wiss. Göttingen, math.–phys. Kl., N. F., Vol. 8, Part 3.
Broom, R. 1932. The mammal-like reptiles of South Africa and the origin of mammals. London, H. F. and G. Witherby.
—— 1933. Evolution—Is there intelligence behind it? South Afr. Jour. Sci., 30: 1–19.
Bulman, O. M. B. 1933. Programme evolution in the graptolites. Biol. Rev., 8: 311–334.
Butler, P. M. 1939. Studies of the mammalian dentition; differentiation of the post-canine dentition. Proc. Zool. Soc. London, Ser. B, 109 (Pt. 1): 1–36.
Camp, Charles L., and Natasha Smith, 1942. Phylogeny and functions of the digital ligaments of the horse. Mem. Univ. California, 13: 69–124.
Clark, A. H. 1930. The new evolution. Zoogenesis. Baltimore, Williams and Wilkins.
Cope, E. P. 1886. The origin of the fittest. New York, Appleton.
Cott, H. B. 1940. Adaptive coloration in animals. London, Methuen and Co.
Crampton, H. E. 1932. Studies on the variation, distribution, and evolution of the genus *Partula*; the species inhabiting Moorea. Pub. Carnegie Inst. Washington, No. 410, pp. 1–335.
Cuénot, L. 1921. La Genèse des espèces animales. Paris, Félix Alcan.
—— 1925. L'Adaptation. Paris, G. Doin.
Darlington, C. D. 1939. The evolution of genetic systems. Cambridge, Cambridge University Press.
Davies, A. M. 1937. Evolution and its modern critics. London, Murby.
Demerec, M. 1933. What is a gene? Jour. Hered., 24: 369–378.
—— 1937. Frequency of spontaneous mutations in certain stocks of *Drosophila melanogaster*. Genetics, 22: 469–478.

Dendy, A. 1912. "Momentum in evolution," in Report (1911) of the Brit. Assoc. Adv. Sci., pp. 277–280.
Dewar, D. 1931. Difficulties of the Evolution Theory. Arnold, London.
Dice, Lee R. 1942. Ecological distribution of *Peromyscus* and *Neotoma* in parts of southern New Mexico. Ecology, 23: 199–208.
Dobzhansky, T. 1937. Genetics and the origin of species. New York, Columbia University Press.
—— 1941. Same. 2d ed.
—— 1938. The raw materials of evolution. Sci. Monthly, 46: 445–449.
—— 1942. Biological adaptation. Sci. Monthly, 55: 391–402.
Dobzhansky, T., and C. C. Tan, 1936. Studies on hybrid sterility. III. A comparison of the gene arrangement in two species, *Drosophila pseudoobscura* and *Drosophila miranda*. Zeit. indukt. Abstamm.-u. Vererblehre., 72: 88–114.
Dollo, L. 1896. Sur la phylogénie des dipneustes. Bull. Soc. Belge Géol. Pal. Hydr., 9: 79–128.
Doutt, J. Kenneth. 1942. A review of the genus *Phoca*. Ann. Carnegie Mus., 29: 61–125.
Dubinin, N. P., and B. N. Sidorov, 1935. The position effect of the hairy gene. Biol. Zhurn. (Moscow), 4: 555–568.
Dubinin, N. P., and others. 1934. Experimental study of the ecogenotypes of *Drosophila melanogaster*. Biol. Zhurn. (Moscow), 3: 166–216.
Dunn, Emmett Reid. 1935. The survival value of specific characters. Copeia, 1935, No. 2, pp. 85–98.
—— 1942. Survival value of varietal characters in snakes. Amer. Nat., 76: 104–109.
Edinger, Tilly. [1941–1942]. Oral reports on equid brain evolution to the Society of Vertebrate Paleontology and the American Society of Mammalogists; not yet published.
Efremov, I. A. 1935. Vipadenie perekhodnikh form v usloviyakh zakhoroneniya drevneyshikh chetveronogikh. [The absence of transitional forms under conditions of fossilization of the earliest tetrapods.] Trav. Inst. Pal. Acad. Sci. U. R. S. S., 4: 281–288. Through the courtesy of the W.P.A., the paper was studied in MS translation made by A. F. Feltyn.
Eimer, G. H. T. 1897. Die Entstehung der Arten. Teil II: Die Orthogenesis der Schmetterlinge. Leipzig, Engelmann.
Eiseley, Loren C. 1942. Post-glacial climatic amelioration and the extinction of *Bison taylori*. Science, N. S., 95: 646–647.
Elton, C. S. 1924. Periodic fluctuations in the number of animals; their causes and effects. Brit. Jour. Exper. Biol., 3: 119–163.
Emerson, S. 1935. The genetic nature of De Vries' mutations in *Oenothera lamarckiana*. Amer. Nat., 69: 545–559.
Eyster, W. H. 1931. Vivipary in maize. Genetics, 16: 574–590.
Fenton, C. L. 1935. Factors of evolution in fossil series. Amer. Nat., 69: 139–173.

Fisher, R. A. 1930. The genetical theory of natural selection. Oxford, Clarendon Press.

Gause, G. F. 1942. The relation of adaptability to adaptation. Quart. Rev. Biol., 17: 99–114.

Gazin, C. Lewis. 1941. The mammalian faunas of the Paleocene of central Utah, with notes on the geology. Proc. U. S. Nat. Mus., 91: 1–53.

George, T. Neville. 1933. Palingenesis and palaeontology. Biol. Rev., 8: 107–135.

Gidley, J. W. 1923. Paleocene primates of the Fort Union, with discussion of relationships of Eocene primates. Proc. U. S. Nat. Mus., 63: 1–38.

Goldschmidt, R. 1938. Physiological genetics. New York and London, McGraw-Hill.

—— 1940. The material basis of evolution. New Haven, Yale Univ. Press.

Goodman, C., and R. D. Evans, 1941. Age measurements by radio-activity. Bull. Geol. Soc. Amer., 52: 491–544.

Gorjanovic-Kramberger, K. 1901. Ueber die Gattung *Valenciennesia* und einige unterpontische Limnaeen. Beitr. Paläont. Geol. Öst.–Ung., 13: 121–140.

—— 1923. Ueber die Bedeutung der Valenciennesiiden in stratigraphischer und genetischer Hinsicht. Paläont. Zeit., 5: 339–344.

Gregory, W. K. 1924. On design in nature, pp. 1–12. Reprinted from Yale Rev., 13: 334–345.

—— 1933. Basic patents in nature. Science, 78: 561–566.

—— 1935a. "Williston's Law" relating to the evolution of skull bones in the vertebrates. Amer. Jour. Phys. Anthrop., 20: 123–152.

—— 1935b. Reduplication in evolution. Quart. Rev. Biol., 10: 272–290.

—— 1935c. The roles of undeviating evolution and transformation in the origin of man. Amer. Nat., 69: 385–404.

—— 1936a. The transformation of organic designs; a review of the origin and deployment of the earlier vertebrates. Biol. Rev., 11: 311–344.

—— 1936b. Habitus factors in the skeleton of fossil and recent mammals. Proc. Amer. Phil. Soc., 76: 429–444.

—— 1936c. On the meaning and limits of irreversibility of evolution. Amer. Nat., 70: 517–528.

—— 1937. Supra-specific variation in nature and in classification; a few examples from mammalian paleontology. Amer. Nat., 71: 268–276.

Haacke, W. 1893. Gestaltung und Vererbung. Leipzig, Weigel.

Haldane, J. B. S. 1932. The causes of evolution. New York and London, Harper.

—— 1942. New paths in genetics. New York and London, Harper.

Hamilton, W. J., Jr. 1939. American mammals. New York and London, McGraw-Hill.

Handlirsch, A. 1909. Ueber Relikte. Verhandl. zool.-bot. Ges. Wien, 59: 183–207.

Hatfield, Donald M. 1940. Animal populations and sunspot cycles. Chicago Nat., 3: 105–110.
Heilmann, G. 1926. The origin of birds. London, Witherby.
Hersh, A. H. 1934. Evolutionary relative growth in the titanotheres. Amer. Nat., 68: 537–561.
Hogben, L. 1933. Nature and nurture. New York, Norton.
Huxley, J. S. 1932. Problems of relative growth. London, Methuen.
Huxley, J. S. (editor). 1940. The new systematics. Oxford, Clarendon Press.
Jacot, A. P. 1932. The status of the genus and the species. Amer. Nat., 66: 346–364.
Kerkis, J. 1936. Chromosome conjugation in hybrids between *Drosophila melanogaster* and *Drosophila simulans*. Amer. Nat., 70: 81–86.
Kinsey, A. C. 1936. The origin of higher categories in *Cynips*. Indiana Univ. Pub., Sci. Ser., No. 4, pp. 1–334.
Kowalevsky, W. 1874. Monographie der Gattung *Anthracotherium* Cuv. und Versuch einer natürlichen Classification der fossilen Hufthiere. Paleontographica, N. F., 2 (No. 3): i–iv, 133–285.
Lotsy, J. P. 1916. Evolution by means of hybridization. The Hague, Martinus Nijhoff.
Lull, Richard Swann. 1917. Organic evolution. New York, Macmillan.
MacLulich, D. A. 1937. Fluctuations in the numbers of the varying hare *(Lepus americanus)*. Univ. Toronto Studies, Biol. Ser., No. 43, pp. 1–136.
Mather, K. 1941. Variation and selection of polygenic characters. Jour. Genetics, 41: 159–193.
Matthew, W. D. 1910. The phylogeny of the Felidae. Bull. Amer. Mus. Nat. Hist., 28: 289–316.
——— 1914. Time ratios in the evolution of mammalian Phyla; a contribution to the problem of the age of the earth. Science, N. S., 40: 232–235.
——— 1915. Climate and evolution. Ann. N. Y. Acad. Sci., 24: 171–318.
——— 1926. The evolution of the horse; a record and its interpretation. Quart. Rev. Biol., 1: 139–185.
——— 1929. On the phylogeny of horses, dogs, and cats. Science, 69: 494–496.
——— 1937. Paleocene faunas of the San Juan Basin, New Mexico. Trans. Amer. Phil. Soc., N. S., 30: i–viii, 1–510.
——— 1939. Climate and evolution. 2d ed., rev. and enl. Spec. Pub. N. Y. Acad. Sci., 1: i–xii, 1–223.
Osborn, Henry Fairfield. 1889. The palaeontological evidence for the transmission of acquired characters. Amer. Nat., 23: 561–566.
——— 1907. Evolution of mammalian molar teeth to and from the triangular type. New York and London, Macmillan.
——— 1915. Origin of single characters as observed in fossil and living animals and plants. Amer. Nat., 49: 193–240.
——— 1925a. The origin of species as revealed by paleontology. Pp. 1–12. Reprinted from Nature, 115: 925, 926, 961–963.

―― 1925b. The origin of species. Part 2. Distinctions between rectigradations and allometrons. Proc. Nat. Acad. Sci., 11: 749–752.

―― 1926a. The origin of species, 1859–1925. Sci. Monthly, 22: 185–192.

―― 1926b. The problem of the origin of species as it appeared to Darwin and as it appears to us to-day. Science, 64: 337–341.

―― 1927. The origin of species; Part 5: speciation and mutation. Amer. Nat., 61: 5–42. [Parts 1–4 of this series are Osborn, 1925a, 1925b, 1926a, and 1926b].

―― 1929. The Titanotheres of Ancient Wyoming, Dakota and Nebraska. U. S. Geol. Surv., Mon. 55, 2 vols.

―― 1932. Biological inductions from the evolution of the Proboscidea. Science, 76: 501–504.

―― 1934. Aristogenesis, the creative principle in the origin of species. Amer. Nat., 68: 193–235.

―― 1936, 1942. Proboscidea. 2 vols. New York, American Museum Press.

―― 1938. Eighteen principles of adaptation in Alloiometrons and Aristogenes. Palaeobiol., 6: 273–302.

Parr, A. E. 1926. Adaptiogenese und Phylogenese; zur Analyse der Anpassungserscheinungen und ihre Entstehung. Abh. Theor. organ. Ent. 1: 1–60.

Pearl, Raymond. 1940. Introduction to medical biometry and statistics. 3d ed. Philadelphia and London, Saunders.

Phleger, Fred B. 1940. Relative growth and vertebrate phylogeny. Amer. Jour. Sci., 238: 643–662.

Phleger, Fred B., Jr., and W. S. Putnam. 1942. Analysis of *Merycoidodon* skulls. Amer. Jour. Sci., 240: 547–566.

Plate, L. 1913. Selektionsprinzip und Probleme der Artbildung. 4 Aufl. Leipzig and Berlin, Verlag von Wilhelm Engelmann.

Quayle, H. J. 1938. The development of resistance in certain scale insects to hydrocyanic acid. Hilgardia, 11: 183–225.

Reinig, W. F. 1935. Über die Bedeutung der individuellen Variabilität für die Entstehung geographischer Rassen. Sitzber. Ges. naturf. Freunde Berlin, Jahrgang, 1935, pp. 50–69.

Rensch, B. 1929. Das Prinzip geographischer Rassenkreise und das Problem der Artbildung. Berlin, Borntraeger.

Robb, R. C. 1935a. A study of mutations in evolution. Part 1: Evolution in the equine skull. Jour. Genetics, 31: 39–46.

―― 1935b. A study of mutations in evolution. Part 2: Ontogeny in the equine skull. Jour. Genetics, 31: 47–52.

―― 1936. A study of mutations in evolution. Part 3: The evolution of the equine foot. Jour. Genetics, 33: 267–273.

―― 1937. A study of mutations in evolution. Part 4: The ontogeny of the equine foot. Jour. Genetics, 34: 477–486.

Robson, G. C., and O. W. Richards, 1936. The variations of animals in nature. London, New York, Toronto, Longmans Green.

Romer, A. S. 1933. Vertebrate paleontology. Chicago, Univ. Chicago Press.
—— 1936. [Review of Säve-Söderbergh, 1935]. Jour. Geol, 44: 534–536.
—— 1941. Man and the vertebrates. 3d. ed. Chicago, Univ. Chicago Press.
—— 1942. Cartilage an embryonic adaptation. Amer. Nat., 76: 394–404.
Rosa, D. 1899. La riduzione progressiva della variabilita a i suoi rapporti coll'estinzione e coll origine delle specie. Torino, Carlo Clausen.
—— 1931. L'Ologénèse; nouvelle théorie de l'évolution et da la distribution géographique des êtres vivants. Paris, Félix Alcan.
Ruedemann, R. 1918. The paleontology of arrested evolution. New York State Mus. Bull., No. 196, pp. 107–134.
—— 1922a. Additional studies of arrested evolution. Proc. Nat. Acad. Sci., 8: 54–55.
—— 1922b. Further notes on the paleontology of arrested evolution. Amer. Nat., 56: 256–272.
Säve-Söderbergh, G. 1934. Some points of view concerning the evolution of the vertebrates and the classification of this group. K. Vet. Akad. Stockholm, Ark. Zool., 26A (No. 17): 20.
—— 1935. On the dermal bones of the head in Labyrinthodont Stegocephalians and primitive reptilia. Medd. om Grønland, Kom. Vid. Unders. Grønland, 98 (No. 3): 1–211.
Schindewolf, O. H. 1936. Paläontologie, Entwicklungslehre und Genetik. Berlin, Borntraeger.
Schlaikjer, E. M. 1935. Contributions to the stratigraphy and paleontology of the Goshen Hole Area, Wyoming. Part 4: New vertebrates and the stratigraphy of the Oligocene and early Miocene. Bull. Mus. Comp. Zool. Harvard, 76: 97–189.
Schuchert, C., and C. O. Dunbar, 1933. A textbook of geology. Part 2: Historical geology. 3d ed. New York, John Wiley and Sons.
Scott, W. B. 1937. A history of land mammals in the Western Hemisphere. Revised ed., rewritten throughout. New York, Macmillan Co.
Scott, W. B., and G. L. Jepsen, 1936. The mammalian fauna of the White River Oligocene. Part 1: Insectivora and carnivora. Trans. Amer. Phil. Soc., N. S., 28: 1–153.
Shull, A. F. 1938. Heredity. 3d ed. New York and London, McGraw–Hill.
Simpson, G. G. 1931. Origin of mammalian faunas as illustrated by that of Florida. Amer. Nat., 65: 258–276.
—— 1935. The first mammals. Quart. Rev. Biol., 10: 154–180.
—— 1936a. Data on the relationships of local and continental mammalian faunas. Jour. Paleont., 10: 410–414.
—— 1936b. Census of Paleocene mammals. Amer. Mus. Novitates, No. 848, pp. 1–15.
—— 1937a. The beginning of the age of mammals. Biol. Rev., 12: 1–47.
—— 1937b. Patterns of phyletic evolution. Bull. Geol. Soc. Amer., 48: 303–314.

―― 1937c. Super-specific variation in nature from the viewpoint of paleontology. Amer. Nat., 71: 236–267.

―― 1937d. Additions to the Upper Paleocene fauna of the Crazy Mountain field. Amer. Mus. Novitates, No. 940, pp. 1–15.

―― 1939. The development of Marsupials in South America. Physis, 14: 373–398.

―― 1940. Mammals and land bridges. Jour. Washington Acad. Sci., 30: 137–163.

―― 1941a. The role of the individual in evolution. Jour. Washington Acad. Sci., 31: 1–20.

―― 1941b. Quantum effects in evolution. Science, 93: 463.

Simpson, G. G., and A. Roe. 1939. Quantitative zoology. New York and London, McGraw–Hill.

Sinclair, W. J., and W. Granger. 1914. Paleocene deposits of the San Juan Basin, New Mexico. Bull. Amer. Mus. Nat. Hist., 33: 297–316.

Snell, G. D. 1931. Inheritance in the house mouse, the linkage relations of short-ear, hairless, and naked. Genetics, 16: 42–74.

Spath, L. F. 1933. The evolution of the Cephalopoda. Biol. Rev., 8: 418–462.

Stadler, L. C. 1932. On the genetic nature of induced mutations in plants. Proc. VI Int. Cong. Genet., 1: 274–294.

Stirton, R. A. 1940. Phylogeny of North American Equidae. Univ. California Pub., Bull. Dept. Geol. Sci., 25: 165–198.

Sturtevant, A. H. 1929. The genetics of *Drosophila simulans*. Pub. Carnegie Inst. Washington, No. 399, pp. 1–62.

Sturtevant, A. H. and T. Dobzhansky. 1936. Geographical distribution and cytology of "sex ratio" in *Drosophila pseudoobscura*. Genetics, 21: 473–490.

Sumner, F. B. 1935. Evidence for the protective value of changeable coloration in fishes. Amer. Nat., 69: 245–266.

Swinnerton, H. H. 1923. Outlines of paleontology. London, Arnold.

―― 1932. Unit characters in fossils. Biol. Rev., 7: 321–335.

Teilhard de Chardin, Pierre. 1941. Réflexions sur le progrès. Peking, privately printed. Pamphlet.

Timofeeff-Ressovsky, N. W. 1933. Rückenmutationen und die Genmutabilität in verschiedenen Richtungen. Parts 3–5. Zeit. Indukt. Abstamm.-u. Vererblehre., 64: 173–175; 65: 278–292; 66: 165–179.

―― 1935. Über geographische Temperaturrassen bei *Drosophila funebris*. Arch. Naturgesch., N. F., 4: 245–257.

―― 1940. Mutations and geographical variation. In Huxley 1940, pp. 73–136.

Truemann, A. E. 1930. Results of recent statistical investigations of invertebrate fossils. Biol. Rev., 5: 296–308.

Turrill, W. B. 1940. Experimental and synthetic plant taxonomy. In Huxley 1940, pp. 47–71.

Vialleton, L., and others. 1927. Le Transformisme. Paris, Vrin.
Villee, Claude A. 1942. The phenomenon of homoeosis. Amer. Nat., 76: 494–506.
Vries, H. de. 1901. Die Mutationstheorie. Leipzig, Veit.
Waagen, W. 1868. Die Formenreihe des *Ammonites subradiatus*. Benecke geog.–pal. Beit., 2: 179–257.
Waddington, C. H. 1939. An introduction to modern genetics. New York, Macmillan.
Watson, D. M. S. 1940. The origin of frogs. Trans. Roy. Soc. Edinburgh, 60 (Pt. 1): 195–231.
Weidenreich, Franz. 1941. The brain and its role in the phylogenetic transformation of the human skull. Trans. Amer. Phil. Soc., n. s., 31: 321–442.
Wheeler, J. F. G. 1942. The discovery of the nemertean *Gorgonorhynchus* and its bearing on evolutionary theory. Amer. Nat., 76: 470–493.
Willis, J. C. 1940. The course of evolution. Cambridge, Cambridge University Press.
Wood, H. E., 2d, and A. E. Wood. 1933. The genetic and phylogenetic significance of the presence of a third upper molar in a modern dog. Amer. Midland Nat., 14: 36–48.
Wright, Sewall. 1931. Evolution in Mendelian populations. Genetics, 16: 97–159.
——— 1932. The roles of mutation, inbreeding, crossbreeding, and selection in evolution. Proc. 6th Int. Cong. Genetics, 1: 356–366.
——— 1935. Evolution in populations in approximate equilibrium. Jour. Genetics, 30: 257–266.
——— 1937. The distribution of gene frequencies in Populations. Proc. Nat. Acad. Sci., 23: 307–320.
——— 1940. The statistical consequences of Mendelian heredity in relation to speciation. In Huxley 1940, pp. 161–183.
——— 1942. Statistical genetics and evolution. Bull. Amer. Math. Soc., 48: 223–246.
Wright, Sewall, Th. Dobzhansky, and W. Hovanitz, 1942. Genetics of natural populations. Part 7: The allelism of lethals in the third chromosome of *Drosophila pseudoobscura*. Genetics, 27: 363–394.
Zimmerman, Elwood C. 1942. Distribution and origin of some eastern oceanic insects. Amer. Nat., 76: 280–307.
Zittel, K. A. von. 1913. Text-book of paleontology. [English revision by C. R. Eastman]. Vol. I. London and New York, Macmillan.
——— 1924. Osnovi paleontologii (Paleozoologiya). Part 1: Bespozvonochnie. [Russian revision by A. N. Ryabinin]. Leningrad. ONTI, NKTP, SSSR.

Index

Abderitinae, 142, 143
Abel, O., 18, 103n, 149, 152, 157, 163, 172
Acceleration, 149
Adaptation, 74 ff., 96; structural and physiological, 64; increasing intensity should accelerate, 80; pictorial representation of relationship between selection, structure, and, 89, 90–91 *graphs;* crucial element in interaction in organism and environment, 180; nature of, 180–83; three sets of distinctions defining, 182; in phyletic evolution, 204 *graph;* and environment, *see* Adaptive zones
Adaptive grid, 191–96 *graphs;* speciation in relation to, 200; relation of phyletic evolution to, 203
Adaptive zones, 188–91; discontinuity between, 189; carnivore and herbivore, 190; capacity for survival proportional to width of, 192; availability of new, 212
Agnatha, 112
Allelomorphs, ratios within a population, 65; random loss of, 66
Ammonites, correlative rates, 13; rates of evolution, 17 *tab.;* cause of apparent saltation in phylum, 100 *graph*
Amphibia, 110; time of origin, 112; primary and secondary trends in evolution of, 168, 169
Anakosmoceras, 15
Ancestry, discontinuity of records concerning, 106, 110
Anchitherium, 93, 103, 114, 162, 210
Anchitheriinae, 93
Antlers, hypertrophy, 171 ff. *passim*
Aonidiella, 64
Apatemyidae, 72, 73 *tab.*
Apes, 125; skull, 166, 167
Archaeohippus, 88, 114, 160, 162
Archaeopteryx, 111
Archaeornis, 111
Aristogenesis, 151
Armadillos, 39, 126, 139
"Arrested evolution," 136
Artiodactyla, estimated duration, 120
Asexual reproduction conducive to slow evolution, 137
Atrypa, 156

Aves, time of origin, 113
Avicula, 132

Basse, E., 135
Bats, 126, 139, 193, 212; evolution of wing, 119
Bifurcation, 204
Biology, experimental defect, xiii
Birds, 110; basic ecological type, 111; earliest known, 112
Bison, 65n
Body size, length of generations correlated with, 63
Bradytelic rates, 133 ff., 147, 148; characteristic of phyletic evolution, 205
Bradytely, in pelecypods, 132 *graph;* factors of, 135–42, 147, 148; large breeding populations, 138; groups usually progressive and high types, 138, 148; lines predominantly adaptive, 140, 148; survival of group, 140; propitious climatic condition, 141; lines almost immortal, 143; phyla in evolutionary state of rest, 149; adaptive grid, 193 *graph,* 194
Brain, in Equidae, 160; in man, 166
Branching accompanied by evolution in opposed directions, 87 ff.; equid, 158; by bifurcation, 204
Breeding populations, 66, 69; increase in mutation rate with decrease in, 70; existence of large, for groups of slow evolution, 138
Brinkman, R., 13, 15, 33, 34, 38; data on ammonites, 100
Broom, R., 35, 71, 152
Browsing, 90 ff., 209
Bulman, O. M. B., 45, 151, 170
"Bundling index," regression of the, 15

Caenolestinae, 142, 143
Caenolestoidea, survival of the unspecialized in, 142 *graph*
Camp, C. L., and N. Smith, 160
Canids, Canidae: adaptive zones, 190
Carinus, 64
Carnivores, Carnivora: procyonid, 20; distribution of genera, 22 ff. *tabs.;* survivorship, 24 ff. *tabs.;* rate of evolution, 29; estimated duration, 120; survivorship

Carnivores, Carnivora (*Continued*)
curves, 127, 129; frequency distribution of rates of evolution in genera of, 128 *graph;* realized and expected age compositions of recent fauna, 130, 131, *graphs;* horotelic rates in, 133
Carnivore zone, 190
Cartilage, selective influence, 169
Casamayoran fauna, 214
Centripetal selection, 83 ff., 96
Cervus elaphus, 177
Cetacea, 212; estimated duration, 120
Chalicotheres, rate of evolution, 17 *tab.*
Character complex rate, 4
Chiton, 88
Chondrichthyes, 112
Chromosomal aberrations, 51, 52, 56
Chromosome mutations, 48
Clark, A. H., 35, 55, 97
Climates, genetic adjustment to, 64
Clone considered as a statistical individual, 183
Cockroaches, 147
Coelacanths, 125
Colbert, Edwin H., xv
Color, adaptation, 181
Condylarthra, 106; ancestral to perissodactyls, 112; estimated duration, 120
Continuity in evolution, 49
Cope, E. D., 114, 152
Cott, H. B., 157
Crampton, H. E., 52
Crocodiles, Crocodilia, 39, 125, 139, 147; trends in, 169
Crossopterygians, 110; ancestral to tetrapods, 111
Cuénot, L., 152, 171
Cuvierian principle of correlation, 164
Cycles, *see* Life cycle; Population cycles

Darlington, C. D., 175, 176
Darwin, C., 75
Davies, A. M., 115n
Deciduous perennials, 64n
Deer, 20; hypertrophy of antlers in *Megoceros,* 171 ff. *passim*
Degenerating structures highly variable, 39, 42, 94
Demerec, M., 155
Dendy, A., 173
Depositional hiatus, apparent saltation in ammonite phylum caused by, 100 *graph*

Determinants of evolution, 30–96; variability, 31–42, 93; mutation rate, 42–48, 94; character of mutations, 48–62, 94; length of generations, 62–65, 94; size of population, 65–74, 95; selection, 74–93, 95; summary, 93–96
Devonian, length of, 112
Dichotomy, 87 ff., 113–114; two patterns of phyletic, 91 *graph*
Didelphis, 127
Differentiation, more rapid and complex for small animals, 63; populations favorable for rapid racial, 67; of large and of small populations, 70; basic, more rapid than later evolution, 121; inadaptive, within small isolated groups, 123; local, of groups within a more widespread population, 199 ff.
Digits, proportions, 6; reduction of, 61, 161
Dimetrodon, 171
Dinosaurs, 212
Direction, modification of without reference to environment, 152; *see also* Trends
Discontinuities of record, minor, 99–105; major systematic, 105–24; mammals, 106 ff., 107 *graph;* attempts to fill gaps by extrapolation, 119; primary cause of deficiency, 121; instable ecological zones, 191 *graph*
Discontinuity in evolution, 49; isolating mechanisms that produce or permit, 97; development of, near level of macroevolution, 98
Dobzhansky, T., 37, 42, 45, 51, 53, 57, 65, 137, 211
—— and C. C. Tan, 57
—— and Sturtevant, 57
Dollo, L., 195
Dormant period in life cycle, 64n
Doutt, J. K., 19
Dragon fauna, 102
Drosophila, survivorship curves, 26 *tab.;* low evolutionary rate, 46$n;* mutant tetraltera in, 52, 53; disadvantageous mutation, 56; chromosome arrangement, 57; forked mutations, 60; mutation of eye color to white, 155
Dubinin, N. P., and others, 58
—— and B. N. Sidorov, 57
Dunbar, C. O., and C. Schuchert, 113
Dunn, E. R., 78, 181

Ecological changes, 121, 122; in new groups: discontinuity of record, 110
Ecological zones, 191
Ectoloph length, growth rate, 9 ff.
Edaphosaurus, 171
Edentata, estimated duration, 120
Edinger, T., 160
Efremov, I. A., 123
Eimer, G. H. T., 150
Eiseley, L. C., 65*n*
Elephants, 145; length of generations: evolution, 63
Elton, C. S., 69
Emergence, epoch of, 113
Emerson, S., 56
Endemism in faunas of isolated Pacific islands, 20
Environment, modification of phenotype, 62; cyclic shifts, 63; genetic adjustment to changes in, 64; equilibrium between tolerance and cyclic change, 140, 148; rectilinear influence on organisms, 150; interaction of organism and, 151; zoning of, 188–91; understood in broadest sense, 188; *see also* Organism and environment
Eocene, duration, 19 *tab.*
Eohippus, 106; transformation into *Equus*, 158
Eohippus-Orohippus, 18
Equidae, structural changes, 8 ff., 10–11 *tabs.;* illustrate basic theorem concerning rates of evolution, 12; occurrence in America of early, 73 *tab.;* factors in subspecific advance, 81; evolution of food habit: browsers and grazers, 90 ff.; true phylogeny and saltations, 103, 104 *graph;* dichotomy, 114; main source of opinion as to rates of evolution, 118; evolutionary trends, 157–64; rates of evolution not constant, 158; phyletic branching, 158; gross size, 159; skull proportions, "preoptic dominance": brain, 160; limb proportions: foot mechanism, 160; reduction of digits, 161; molarization of premolars: hypsodonty and cement: molar pattern, 162; broader trends were adaptations effective in wide range, 169; European, 195; phases of history interpreted as quantum evolution, 208 *graph*
Equus, new mutations in ancestry, 45; teeth, 59; Idaho deposit, 72; transformation of *Eohippus* (*Hyracotherium*) into, 158
Equus caballus, 5
Eusmilus, 196
Evans, R. D., and C. Goodman, 24
Evolution, effect of increasing specialization in studies, xi; dominant factor of, 35; during generations, 62; between generations, 62, 94; optimum condition for rapid, 67; effect of slight selective advantages upon, 81 f.; explosive stages, 89, 139, 213 *graph;* unusually rapid, tends to be self-limiting, 149; metaphysical tendency to proceed in straight lines, 152; pattern of step-like, 194 *graph*
Expansion, periods of, accompanied by high mutation rates, 45
Explosive stages of evolution, 89, 139; by multiple quantum steps into varied adaptive zones, 213 *graph*
Extinction, differential, at close of Pleistocene, 64; random gene, 69; numerical estimates of size of extinct populations, 71, 95; liability to, proportional to rate of evolution, 143; most general cause, 176; secondary causes, 179
Eyster, W. H., 64*n*

Fecundity tends to promote rapid evolution, 138
Felidae, Felids, 195; growth rates, 6; discontinuities, 190
Felines, 196; discontinuities between machairodontines and, 190
Felis, 144
Fenton, C. L., 46*n*, 123, 156
Fisher, R. A., 36, 55, 58, 65 ff. *passim*, 80, 81, 157, 173
Fissipedes, 190
Flight, loss of, in birds, 110
Flight structure of birds in basic ecological type, 111
Foot mechanism, 160
Fossils, numerical estimates of, 71; reveal only small proportion of species that have lived, 104
Fox, Arctic, 69
Functional integration, 56

Gaps in record, *see* Discontinuities of record
Gastropods, 88
Gause, G. F., 183
Gazin, C. L., 102

Gene mutations, 48, 50 ff.; multiple and systemic effects, 51
Genera, duration within a phylum, 16, 20; as units for determining organism rates, 16; sequence of successive, 17; fossil record, 20; continuity or intergradation between species and, 58; persistent types: immortal types, 136
Generations, length of, 62–65, 94; correlated with size, 63
Genetics, synthesis of paleontology and, xi; population genetics, xii; defect, xiii; direct study of change impossible to paleontologist, 3; Mendelian, 48; simplest pattern, 65, 95; preadaptation consistent with data of experimental, 76; effect of selection upon a gene or genetic system, 78 ff. *passim*
Genotypes in successive generations, 40
Geographical relicts, 144, 145
Geographic distribution, most groups first expand and then contract in, 145, 146 *chart*, 148
Geographic range, length of generations correlated with, 63
Geological ages, translation into terms of time, 24; duration of period, 112n
George, T. N., 171
Germann, John C., xv
Gidley, J. W., 119
Giraffes, 145
Goldschmidt, R., 49 ff. *passim*, 97, 98, 116, 166
Goodman, C., and R. D. Evans, 24
Gorgonorhynchus, 54
Gorjanovic-Kramberger, K., 134
Granger, W., and W. J. Sinclair, 102
Graptolites, periods of increase and decline, 45; effect of extreme lobation, 170
Grazing, adaptation, 90 ff., 208, 209
Gregory, W. K., 151; quoted, 164
Grid method, *see* Adaptive grid
Groups, *see* Populations
Growth rates, 4, 6
Gryphaea, 133, 170, 171 *tab.*, 174

Haacke, W., 150
Hair texture in man, 166
Haldane, J. B. S., 36, 65, 67, 69, 81, 82, 152, 171 ff. *passim*, 181
Hamilton, W. J., 69
Handlirsch, A., 144
Hare, variation, 69
Hatfield, D. M., 70

Heilmann, G., 111
Henricosbornia, 45
Henricosbornia lophiodonta, 32
Herbivore zone, 190
Hersh, A. H., 79
Hereditary variation, *see* Variation, genotypic
Heredity, *see* Inheritance
Hibernation, 64n
Hipparion, 88, 103, 161
Hipparion gracile, 134
Hogben, L., 62
Homoeosis, 52 ff.
Horotelic evolution, rates, 133 ff., 147, 205; slow phyla live longer than fast, 143, 148; typical situation, 193 *graph*, 194
Horse, skull proportions, 4 ff.; relative rates in, 5 *tab.*; digit proportions, 6; hypsodonty, 7 ff.; samples of fossil, 8 ff.; measurements on M^3, 9 *tab.*; rates of evolution, 17 *tab.*, 19, 39, 145; changes in rates, 18 *tab.*; variation of paracone height and ectoloph length, 38; morphological evolution, 45; true equine, 59; tooth characters, 59; reduction of lateral digit, 61; length of generations, 63; evolution, 63, 90 ff., 103, 106, 118, 157; dichotomy, 88; evolution of food habit: browsing, grazing, 90 ff., 92 *illus.*; of early Pliocene, 134; forest and desert, 163; postadaptation before limb was evolved, 187
Hovanitz, W., S. Wright, and T. Dobzhansky, 42
Huxley, J. S., 4, 177
Hybridization, 35
Hypohippus, 63, 87, 103, 162; discontinuity between *Merychippus* and, 98
Hypohippus-Megahippus, line, 163
Hypohippus osborni, 8 ff.
Hyracotherium, 87, 93, 103, 106; structural gap between *Condylarthra* and, 106; transformation into *Equus* (*q. v.*), 158; size, 159
Hyracotherium borealis, 8 ff.
Hyracotherium-Equus line, 4; genera, 17; number of generations, 45; change in tooth structure, 45
Hyracotherium-Hypohippus line, 162
Hyracotherium-Mesohippus line, brain transformation: foot mechanism, 160; molarization of premolars, 162
Hypsodonty, *see under* Teeth

Ichthyosaurs, 212
Inadaptation, 80
Inadaptive characters spread in populations, 181
Inertia, trend, and momentum, 149–79; evolutionary analogue of, 150; law of, exemplified by evolution of horse, 157; in sense of lag, 176, 179
Inheritance, transmission of inherited characters, xii; particulate or blending, 49; possible combinations of discrete hereditary units into individual zygotes, 80; linearity inherent in conservative effects of, 154, 178; hereditary factors and natural selection, 174; *see also* Variation, genotypic
Insectivora, estimated duration, 120
Inversion, 57
Invertebrates, length of generations: evolution, 63; size of earliest, 110; why fossilization is improbable, 116
Islands, evolution on isolated, 19, 20

Jepsen, G. L., and W. B. Scott, 98

Kangaroos, 39
Kinsey, A. C., 98
Koala, 140
Kosmoceras, correlation and regression, 13, 14 *tab.*, 100; transformation of species, 33; variability, 38
Kowalevsky, W., 103

Lag, 176, 179
Lagomorpha, estimated duration, 120
Lamarckian theory, 75, 76
Land animals, return to sea, 110
Latimeria, 112, 125, 138
"Law of inertia," 157
Law of progressive reduction of variability, 31, 32, 36
Leda, 132
Lemuroids, 125
Leptonychotes weddelli, 187
Life cycle, relationship between cyclic or secular environment and, 63, 94; adaptations in different stages, 64; life span, co-ordinations with climatic cycles, 64n; long, conducive to slow evolution, 137
Lima, 132
Limb proportions, 160
Limnaea, 134
Limulids, 125, 138

Limulus, 144
Linear selection, 83, 85, 89, 96
Lingulids, 125, 138
Litolestes notissimus, cingulum on lower cheek teeth, 60 *tab.*, 61
Litopterna, estimated duration, 120
Lizards, 39
Lotsy, J. P., 35
Lull, R. S., 151
Lung fishes, 195
Lutrinae, 186
Lynn, Ida M., xv

Machairodonts, 176; canines, 172; extinction, 196; discontinuities between felines and, 190; adaptive zone, 190, 191, 196
MacLulich, D. A., 69
Macro-evolution, xiii, 52, 57, 97–124
Mammalia, time of origin, 112; group rate for, 127
Mammalian orders, inadequacy of record, 106 ff., 107 *illus.*; available records of ancestry, 108–9 *tab.*; size of earliest members compared with later, 110; that arose in epochs of emergence and orogeny, 113; estimated durations, 120 *tab.*
Mammoths, 172
Man, variation in, 40; rate of evolution, 125; brain, 166; skull, 166, 167; hair texture: skin color, 166; fundamental character that distinguishes from apes, 167
Marsupials, Marsupialia: ancestry, 111, 127; estimated duration, 120
Mastodonts, geographic distribution, 146 *chart*
Mather, K., 36
Matthew, W. D., 15 ff. *passim*, 50, 58, 99, 102, 106, 114, 195n; quoted, 49, 176
Mechanical motion, application of laws of, to interpretation of evolution, xii, 149
Mega-evolution, 105; coincidence between tectonic episodes and rise of new taxonomic groups, 113; saltation theory of, 115; conditions that lead to, 122, 123; typical process, 124
Megoceros, hypertrophy of antlers, 171 ff. *passim*
Mendelian genetics, 48
Mendelian mutation, 76
Merychippus, 88, 114; hypsodonty and cement, 162; rapid evolution into grazing forms, 210
Merychippus-Equus line, 162

Merychippus-Hypohippus line, discontinuity, 98
Merychippus paniensis, 8 ff.
Merychippus-Pliohippus transition, reduction of digits, 161
Mesohippus, 59
Mesohippus bairdi, 8 ff.
Mesohippus-Merychippus transition, foot mechanism, 160
Metaphysical speculation, 76, 77
Mice, "short ear" in house mouse, 51; death rate of melanos, 181
"Micro" and "macro," conflict of opinion re size implications of, 97 ff.
Micro-evolution, xiii, 57, 97–124
Miocene, duration, 19 *tab.*
Miohippus, 59, 88, 114, 119
Mode, problems suggested by word, xiv, 3
Modes of evolution, 197–217 *graphs;* effects of mutation rates, 42 ff, 94; major, 198; speciation (*q.v.*), 199–202; phyletic (*q.v.*), 198, 202–06; quantum (*q.v.*), 206–17; characteristics of main modes, 216
Modiola, 132
Mollusks, 24, 88
Momentum, inertia, trend, and, 149–79; principle of, 150, 177; evolutionary, 170–77, 178
Momentum effects, 172, 174, 175 *graph,* 178
Monkeys, 125
Monograptids, 170
Morphogenesis, xiv
Morphological change, 16; amount of, relative to a standard, 3
Morphological characters, classification of, 166
Morphological differentiation, occurrence of subspecific, 19
Mustelidae, fission of, into Lutrinae and Mustelinae, 186
Mustelinae, 186
Mutation rate, 42–48, 94; effects on tempo and mode of evolution, 42 ff., 94; might become a primary factor of evolution, 47; length of generations a factor in, 62; expressed per individual, 62; relationship to population size, 66, 68, 95; relationship of selection rate to, 67; affected by natural radiation, 70; increase in, with decrease in breeding population, 70; coincidence between physical events and evolutionary acceleration, 122; low, conducive to slow evolution, 137
Mutations, number of, a determinant of variability, 41; size and nature of, 44; new in ancestry of *Equus,* 45; character, 48–62; use of word, 49; size of, 50, 94; phenotypic effects, 50; systemic, 51, 52; appearance of a mutant individual not evolution, 53; most possibilities of, rigidly limited, 55; large, less frequent than small: less likely to be advantageous, 58; exception to rule of origin of new characters, 61; new types of organisms arise at random by Mendelian, 76; random, 77; supplies materials of creation, 80; disadvantageous, 141, 148; directional, 154, 155, 178; *see also* Saltations

Nannippus, 88, 160
Natural selection, elements, 30; an eliminating force, 31; hereditary factors that have little bearing on: those that will be favored by, 174; basis for theory of, 180
Neohipparion, 63
Neohipparion occidentale, 8 ff.
Neo-Lamarckian theories, 75, 95
Neomenia, 88
Notoungulates, estimated duration, 120; "explosive" evolution, 140
Nucula, 132
Numerical relicts, 144

Old age, racial, 214
Oligocene, duration, 19 *tab.*
Ontogeny, 4*n*
Opossums, 39, 125, 139, 147; length of generations: evolution, 63; possibly ancestral to marsupials, 111, 112; among most successfully adapted of all mammalian types, 140
Oreodonts, 6
Organism and environment, 180–96; interaction of, 151; relationships that bear most directly on tempo and mode, 180; nature of adaptation, 180–83; real and prospective functions, 183–86, 184–85 *graphs;* preadaptation and postadaptation, 186–88; adaptive zones, 188–91; adaptive grid, 191–96 *graphs*
Organisms, rate of evolution for whole, 4, 15–20; sudden rise of new types, 48 (*see also* Mutations); new random, 76; in-

Organisms *(Continued)*
  adaptative, 78; chances of discovering remains of extinct, 116
Orogenic epochs, 113
Orohippus, size, 159
Orohippus-Epihippus, 18
Orthevolution, 150
Orthogenesis, 48, 150 ff., 177, 194; most widely cited examples of, 157; picture of horse evolution different from most ideas of, 163; relation to extinction of monograptids, 170
Orthoselection, 150, 152; trends of molar patterns in horses, 163; rectilinear sequences most consistent with theory of, as primary factor, 178
Osborn, H. F., 43, 44, 48, 49, 59, 78, 79, 99, 114, 150, 151, 152, 157, 203, 206
Osteichthyes, 112
Ostrea, 132, 133; progressive curvature in shell from a primitive *Ostrea* to advanced *Gryphaea*, 170, 171 *tab.*

Palaeothentinae, 142, 143
Paleontological record, incompleteness of, 104; conditions that have influenced, 116; rectilinearity, 152; discontinuities in, *see* Discontinuities of record
Paleontologists, rise of new characters, 49, 55; basis of principles and theories of evolution advanced by, 73; opinion re size implications of "micro" and "macro," 97 ff.
Paleontology, synthesis of genetics and, xi
Paleotherium, 103
Panda, 140
Pantodonta, estimated duration, 120
Paracone height, growth rate, 11
Parahippus, 59, 88, 114, 162
Parahippus-Merychippus, 18
Paramecia, experiments on adaptation and acclimatization of, 183
Parr, A. E., 184, 186
Partula, new races of, 52; sinistral mutations, 60
Pearl, R., 24, 40
Pearsonian coefficient of variation, 33
Pelecypods, Pelecypoda: distribution of genera, 21 *tab.*; survivorship, 24 ff. *graphs, tab.*; expected and realized survival of genera of various ages, 130 *tab.*, 131 *graph;* bradytely, 132 *graph;* present fauna result of different rate distributions, 133; group rate: survivorship curves, 127; frequency distribution of rates of evolution in genera of, 128 *graph*
Penguins, 189
Perissodactyls, Perissodactyla: rate of evolution, 17 *tab.*; ancestry, 106, 112; estimated duration, 120; premolars, 162
Persistence, *see* Evolution, slow
Phenacodus, tooth cusp, 43
Phenogenetics, xii
Phenotype, genetic factors determinants of evolution of, 3; changes in, 48; characters produced by single mutants, 52; environmental factor may modify, 62; characters tend to vary together, or be associated, or show functional correlation, 164
Phleger, F. B., 6
—— and W. S. Putnam, 6, 7
Phoca vitulina, 19
Phyla, effort to divide into generic and specific stages, 16; duration of genera within a phylum, 20; only mode of evolution within the major, 35; trend toward larger size, 86; in evolutionary state of rest, 149; tend to evolve in one direction, 150, 177; modal progression in, 194
Phyletic evolution, 198 *graph*, 202–06, 204 *graph*
Phyletic rate, 4
Phylogenetic relicts, 144, 145
Phylogeny, a chief aim of research, xiv; variation in a branching, 33, 34 *tab.*; felid, 195
Pinnipede, adaptive zone, 190
Placodermi, 112
Plate, L., 150 ff., *passim*
Plesippus, 72
Pliocene, duration, 19 *tab.*
Pliohippus-Equus transition, 103*n*
Polydolopidae, 142
Polyploidy, 56
Polytomy, 114
Population cycles, 65 ff.; relationship between cyclic or secular shifts in environment and, 63, 95
Population genetics, xii
Populations, group rates, 4; group rates and survivorship, 20–29; interbreeding, the essential unit in evolution, 31; new groups represent major ecological adaptive change, 110; appearances of new

Populations (*Continued*)
  groups with unrecorded origin-sequences, 112
  —— large: differentiated into local groups, 70; fragmented into numerous small isolated lines, 123
  —— small: inadaptive and liable to extinction, 70, 95; capable of most rapid evolution, 71; connection between rapidity of evolution and, 118 *chart,* 123; ancestral, splitting into small groups most likely to produce new types, 122
Population size, 65–74; length of generations correlated with, 63; as determinant of rates and patterns of evolution, 66; conditions favorable for evolution, 67, 95; influence on maximum rates of evolution, 68 ff.; a primary factor in bradytely, 138
Postadaptation, 186–88
Preadaptation, 36, 76, 79, 80, 95, 184, 186–88, 210
Proboscideans, 47, 48, 53, 114; estimated duration, 120
Procyonid carnivores, 20
"Programme-evolution," 170
Progressive pre-optic predominance, 4
Pteria, 132
Pterodactyls, 193, 212
Puerco faunas of New Mexico, 101; fauna in Utah intermediate between Torrejon and, 102
Puma, 144
Putnam, W. S., and F. B. Phleger, 6

Quantum evolution, 199, 206–17; three phases of: interval between two equilibria, 207; deficiency of documentation, 209; populations in intermediate condition unstable, 210; situation ideal for: rise of new organic type, 216
Quayle, H. J., 64

Rabbits, 126
Racial old age, 214
Racial vitality, 215
Raciation, 199
Radiation phenomena, population cycles may be correlated with, 70
Rates of evolution, xiii, 3–29; kinds, 3 ff.; relative, in genetically related unit characters, 4–7; relative and absolute, in genetically independent characters, 7–12; correlative, 12–15; organism rates, 15–20; group rates and survivorship, 20–29; differences within same group at different times, 28; inversely proportional to variability, 40, 94; low, inconsistent with normal mutation rates, 46; length of generations influenced by, 62–65, 94; result of mutation rate on, 62; influence of intensity of selection, 82, 83, 96; coincidence of increased, with decreased size of population, 118 *graph,* 123; fluctuations, 119; usually rapid in major unrecorded periods, 119; in systematic gaps, 120; basic differentiation and later adjustment, spread, and diversification, 121; low-rate and high-rate lines, 125–48; medium-rate line, 125; distributions, 126–35, 147; tend to accelerate, 143; liability to extinction proportional to, 143
—— slow: three distinguishable phenomena, 136; individual characters associated with, 137; large breeding populations, 138, 148
Readaptation, 80
Recombination, 55
Record, *see* Paleontological record
Rectigradation, 151
Rectilinear evolution, 150–57, 177–78; some degree of rectilinearity common in evolution, 152; not universal, 152, 177; selection versus internal control, 154; typical of large populations evolving at moderate rates, 153, 177; typical situation in horotelic evolution, 194; *see also* Orthogenesis
Relicts, 144–47
Reptiles, Reptilia: estimate of populations in Karroo, 71; ancestry, 110; ancestral to birds, 112; time of origin, 112; orders that arose in epochs of emergence and orogeny, 113; early, divided into Mammalia and later Reptilia, 114; elongation of dorsal spines in Permian, 171; mammal-like, 216
Rest, evolutionary state of, *see* Inertia
Reversion in phyletic evolution, 204
Revolutions, no vertebrate classes originated during, 112; names: greatest coincidence of, with extinction of old groups, migrations, and expansion, 113; tend to weaken archaic lines, 123
Robb, R. C., 4 ff. *passim,* 61, 79

Index

Robson, G. C., and O. W. Richards, 77, 79, 157
Rodents, isolated, 19; cricetine, 20; first appearance, 110; estimated duration, 120
Roe, Anne, xv
Romer, A. S., 113, 169
Rosa, D., 31, 36, 116, 139, 144
Ruedemann, R., 46, 136, 137, 139
Ruminants, premolars, 162

Saltations, 49, 94; commonly chromosomal aberrations, 56; as a normal mode of evolution, 57; new taxonomic units arise by, 52; in ammonite phylum, 100 *illus.*; causes, 105; saltation theory of megaevolution, 115
Säve-Söderbergh, G., 113
Schindewolf, O. H., 58n, 99
Schlaikjer, E. M., 59
Schuchert, C., and C. O. Dunbar, 24, 113
Scott, W. B., 195n
Scott, W. B., and G. L. Jepsen, 98
Seal, marine, isolated in fresh-water lake, 19
Sebecus Simpson, 169
Sedimentation, variation in rate of, 14
Segregation, 55; of variants, 33
Selection, 74–93, 95–96; relationship of population size to, 66; rate relative to mutation rate, 67; role of, 74–80; in small populations, 78; effect upon genetic combinations and arrangements, 78; a most essential factor in evolution, 80, 96; intensity, 80–83, 96; effect upon variability, 81 ff.; direction, 83–93, 96; pictorial representation of relationship between structure, adaptation, and, 89, 90–91 *illus.*; selective value of incipient character, 156; on young favoring characters disadvantageous to old, 172, 178; a factor in evolution of teeth, 187n; a conditioning factor throughout preadaptation and postadaptation, 188; *see also* Natural selection
Selection landscapes, 89, 90–91 *illus.*
Selection vectors, 83
Senile variability, 214
Sharks, 185
Shells, 88
"Short ear" in house mouse, 51
Sidorov, B. N., and N. P. Dubinin, 57
Simpson, G. G., 61, 64, 142, 144

Sinclair, W. J., and W. Granger, 102
Sirenia, estimated duration, 120
Size, correlated with length of generations 63; tendency of optimum for, to shift toward larger size, 85; abrupt increase, 86; can vary only in two directions, 88; small, contributes to rapidity of evolution and to paucity of fossil discoveries, 121; *see also* Population size
Skin color in man, 166
Skull, in horse, 4 ff.; in Equidae, 160; in apes and men, 166, 167
Sloths, 39
Slugs, 88
Smilodon, 171, 176
Smith, N., and C. L. Camp, 160
Snails, 88
Snell, G. D., 51
Spath, L. F., 99
Speciation, 198, 199–202, 200 *graph;* mutation a disturbance of regular course of, 44; always adaptive, 78; direction in, as a mode, 200; phenotypic differences involved, 201; results of accumulation of changes of mode, 202
Species, new, 52; continuity or intergradation between genera and, 58; successive, 202
Sphenodon, 125, 138, 139, 145
Spiritual values, effect of emphasis upon, 76
Stadler, L. C., 51
Stirton, R. A., 17, 59, 114
Structure, disadvantageous and advantageous characters, 78 f.; pictorial representation of relationship between selection, adaptation, and, 89, 90–91 *graphs;* correspondence of time intervals with breaks in, 115
Stufenreihe, 194 *graph*
Sturtevant, A. H., and T. Dobzhansky, 57
Sumner, F. B., 157
Sunspot cycles, 70
Survivorship, group rates and, 20–29; standard pattern of, 127; correlation with rate of evolution, 127; expected and realized, 129, 130 *tab.;* circumstances involved, 188; adaptive grid of conditions leading to, 192 *graph;* of the specialized, 192; of the unspecialized, 193
Survivorship curves, 24
Swinnerton, H. H., 48, 170

Tachytelic lines, change from one zone to another, 193 *graph*, 194; quickly become extinct or cease to be tachytelic, 143; rates of evolution, 134 ff., 147
Taeniodonta, estimated duration, 120
Tan, C. C., and T. Dobzhansky, 57
Tapirs, 39, 106
Taxonomic relicts, 144
Taxonomic system, a rich source of data re organism rates, 16
Teeth, of horse, 7 ff., 59; rate of evolution of equid molars, 10 *tab.;* hypsodont, 7 ff., 16, 86, 90, 119, 209; variation of paracone height and ectolph length, 38; cusp representing a new unit mutation, 43; dentition, 56; crochet mutation, 59, 60; incipient united metaloph, 60; brachyodont, 86, 90; durability of, and increase in size, 87; molarization of premolars: hypsodonty and cement: molar pattern in Equidae, 162; evolution of, in *Leptonychotes*, 187
Teilhard de Chardin, P., 152
Tempo, problems suggested by, xiv, 3; effects of mutation rates on, 42 ff., 94
Tertiary epochs, estimates of durations, 18, 19 *tab.*
Tetrapods, ancestry, 111
Thylacines, zone, 190
Time intervals, correspondence with structural breaks, 115 (*see also* Discontinuities of record)
Timofeeff-Ressovsky, N. W., 56, 58, 155
Titanotheres, 106, 114, 171
Tolerance, physiological, 64; equilibrium between, and cyclic environmental change, 140, 148
Torrejon faunas of New Mexico, 101; fauna in Utah intermediate between Puerco and, 102
Trägheitsgesetz, 157
Transitional forms, rarity of, as fossils, 123
Transitional stages, 210; gaps in record of major, 115
Translocation, 57
Trends, inertia, and momentum, 149–79; uni-directional nature, 154; in Equidae, 157–64; reversals of direction, 161; primary and secondary, 164–69, 178; simultaneous, 164; existence of, usually clearly discernible, 203; *see also* Direction
Trueman, A. E., 170, 171
Turrill, W. B., 62

Turtles, 126
Types of evolution, 97–124
Typotheres, 110

Ungulates, evolutionary history, 103; common ancestry, 106; development of adaptive specializations, 110; occurrence and interplay of major modes of evolution, 213 *graph*, 214
Unit character rate, 3
Unspecialized, survival of the, 142–43, 148

Valenciennesia, 134
Variability, 30–42, 93; relationship of, and evolution, 30; law of progressive reduction of, 31, 32, 36; application of term, 31; high, 35; of degenerating characters, 39; in living groups having fast or slow rates, 39; in man, 40; rate of evolution inversely proportional to, 41, 94; and mutation rate, 44; effect of intensity of selection, 81, 82; senile, 214
—— intragroup: segregation or selection of, 41; proportional to size of breeding population, 67
Variables, ratios of a number of, in different sorts of populations set up as models, 65
Variants, unequal distribution of ancestral, among descendant groups, 32; segregation of, 33
Variation, paleontologist acquiring new attitude toward, xii; genotypic and phenotypic, 31; application of term, 31; continuous preservation of, in all directions, 36 *tab.;* coefficient for paracone height and ectoloph length in horses, 38 *tab.;* analogous variates in widely disparate mammals, 38; homologous variates, 39; *see also* Saltation
—— intragroup and intergroup, 31, 50, 52; transformation, 34; diversity, 45; distinction between, in rate of evolution, 127; intergroup variation induced by onset of unfavorable conditions, 215
Vertebrata, length of generations: evolution, 63; size of earliest, 110; no classes shown to have originated at time of a revolution, 112; tendency of groups toward larger size, 159
Vertebrata, new arrangement of, 113
Vialleton, L., 152
Villee, C. A., 52, 53
Vitalistic theory, 76, 95

Viverrids, 196
Vries, de, H., 48, 49, 56

Waagen, W., 48, 49
Waddington, C. H., 51, 154, 172, 174
Watson, D. M. S., 168, 169
Weddell's seal, 187
Weidenreich, F., 151, 166
Whales, 126, 185
Wheeler, J. F. G, 54
Willis, J. C., 144, 157, 163

Wood, H. E., and A. E. Wood, 56
Wright, S., 36, 42, 65 ff. *passim,* 80, 81, 82, 89, 152, 156

X-rays, affect mutation rates, 70

Zimmerman, E. C., 20
Zittel, K. A. von, 24
Zugokosmoceras, 14 *tab.,* 15, 100
Zygotes, possible combinations of discrete hereditary units into individuals, 80